Ecuadorean Agrofoi

H. Borgtoft Pedersen & H. Balslev

AAU REPORTS 23

Botanical Institute Aarhus University 1990

this issue in collaboration with

Pontificia Universidad Católica del Ecuador, Quito

Contents

Preface

The first six chapters of this book introduce the reader to some basic concepts relating to the utilization of palms. The next five chapters treat five palm species which we have studied in Ecuador; we review what has been published about these five species in Ecuador and elsewhere, we then present our own observations on their ecology, biology, and utilization, and finally we discuss the potential they represent for being used in different kinds of land-use systems

Our aim is twofold. First, we wish to document some of the value which lies in the natural vegetation of the tropical forests of Ecuador; second, to present the information about these Ecuadorean palms in a way that can be used. We hope this book will be useful for conservation efforts, and for sustainable exploitation of the natural vegetation resources in Ecuador.

We describe extractivism based on wild stands of palms, and the use of palms in agroforestry systems. Agroforestry involves the cultivation of palms, but in land-use systems where they are mixed with other crop species, and sometimes with animal components as well. Palm plantations is a third method described. It involves usually large stands with a single species and is most relevant for industrial utilization of palms. Finally we give a brief description of the use of palms as ornamentals, and discuss a number of different palm-harvesting methods.

The five species, which we have selected here to demonstrate the economic potential of Ecuadorean palms, represent different ecological types, and they are also very different in terms of how much they have penetrated the Ecuadorean cash economy. They grow on soils which vary from nutrient poor to rich, and which may or may not be inundated. Their economic importance, at present, varies from being used very locally only to being exported from the country. The five species and their parameters are as follows:

1. *Astrocaryum jauari* grows in seasonally inundated areas and no products from it are marketed in Ecuador.

2. *Mauritia flexuosa* grows in permanently water logged soils along the rivers and in the back-swamps. Only rarely do products from it find their way to the market places.

3. *Jessenia bataua* is very common in forests on *terra firme* soils, but it may also be found in nutrient poor swamps. Fruits and oil extracted from it are commonly found on the local markets.

4. *Ammandra natalia* has, so far, only been found growing on *terra firme* soils. The fibers from its leaf sheaths are marketed throughout the country.

5. *Euterpe chaunostachys* is a common plant in the nutrient rich swamps along the Pacific coast of Ecuador. Canned palm hearts, produced from its crown-shaft, are the basis for a small industry which markets this product within Ecuador, but also on the international market.

A final chapter includes a number of diagrams that show the climatic conditions where the palms occur naturally. These diagrams are intended as a help to those who wish to experiment with growing the palms.

A list of addresses for research institutions is included to facilitate possible contacts to those researchers already involved in cultivation and exploitation experiments for these five palm species.

Henrik Borgtoft Pedersen and *Henrik Balslev*
Aarhus, August 1990

Borgtoft Pedersen & Balslev

Acknowledgements

We are indebted to many persons who helped us during our fieldwork in Ecuador. Tjitte de Vries, Laura Arcos Terán, and Peter Møller Jørgensen provided working facilities at Pontificia Universidad Católica del Ecuador in Quito. Mr. V. Chalá and others from the INIAP station at Payamino helped and provided facilities there. E. Ramón, H. Carreno, R. Naranjo, B. Bergmann, A. Barfod, L. P. Kvist, and U. Blicher-Mathiesen helped in the field and commented on the work.

We are most grateful for the help and hospitality offered by the staff of the herbaria that we visited (BH, COL, F, FTG, MO, NY, US) while collecting information for this study. We thank V. K. S. Shukla and S. Nielsen (Aarhus Olie) and P. Fromholdt (Grindsted Products A/S) for analyses of composition of palm products, and J. M. Olesen for identification of insects. We thank the Ministerio de Agricultura y Ganadería in Quito, especially Sergio Figueroa, Arturo Ponce, and Angel Lobato for research permits. A special thank go to Dennis V. Johnson for several suggestions and corrections to the manuscript.

Financial support from Danish Natural Science Research Council, the Danish International Development Agency (Danida), The Danish Ministry for Education, and Etatsraad C. G. Filtenborg og hustru Marie Filtenborgs Studielegat is gratefully acknowledged.

Authors

Henrik Borgtoft Pedersen. *Born 1959. Cand. scient. Aarhus University 1988. Since 1989 director of Herbario QCA, Departamento de Ciencias Biológicas, Pontificia Universidad Católica del Ecuador, Apart. 2184, Quito, Ecuador.*

Henrik Balslev. *Born 1951. Cand. scient. Aarhus University 1978. PhD City University of New York and New York Botanical Garden 1982. 1981-1984 director of Herbario QCA, PUCE, Quito. Since 1984 associate professor and curator of the herbarium, Botanical Institute, Aarhus University, Bygn. 137, Universitetsparken, DK- 8000 Aarhus C., Denmark.*

Ecuador is located on the Pacific coast of the South American continent. Enclosed by Colombia to the north and Peru to the south it spans equator from 1°30' north to ca. 5° south latitude. The Andean cordillera divides the country into three distinct natural regions. One is the *coastal plain* towards the Pacific Ocean, the second is the *Andean highlands*, and the third is *Amazonas* to the east on the western rim of the great Amazon basin. This map shows the location of the places mentioned in this text.

1. INTRODUCTION

Palms are beautiful and conspicuous in the vegetation throughout Ecuador, but they are more than that. Indigenous people have known for centuries how to exploit palms because they provide so many of their daily needs: food, beverage, medicine, fiber, thatch, wood for construction, hunting gear, and a variety of tools. In addition palms produce large quantities of fruits that are consumed by mammals, birds, and fish which, in turn, are important sources of protein to humans. Palm exploitation based on indigenous knowledge yields a number of products of commercial value, such as palm heart, vegetable ivory, oil seeds, and fibers.

Because of their many useful features wild palms constitute an important resource for the development of sustainable agricultural systems. The use of wild palm species can contribute to the development of land-use systems that are ecologically and socially adapted to the natural and cultural conditions in Ecuador. There are a number of reasons for this: palms are already present; they are common, or can be grown, on land unsuitable for other crops; they are perennials that make more stable systems than annuals; single species often provide cash crops as well as subsistence crops; and, finally, the value of wild palm species is widely known by indigenous people and settlers.

The richness of palm products is in part a reflection of the high number of palm species growing in Ecuador. Up until now we have recorded 129 native species in 34 genera and 14 introduced species in 11 genera of palms in Ecuador (Balslev and Barfod 1987). Worldwide the palm family has about 2800 species in 200 genera, of which 837 species in 81 genera occur in South America (Uhl and Dransfield 1987, Moore 1973a).

Rapid forest destruction can be seen in almost every part of Ecuador. Palms, however, often escape the axe and are left standing in pastures because they are useful and have ornamental value, and because they produce little shade for pastures and crop plants. They are resistant to fire, and their hard trunks are often difficult to cut. Since the palms are not cut down they are often abundant on cultivated lands, and this gives the impression that they are not as threatened by deforestation as are other kinds of plants. This is not true, however. It is only a matter of time before they will disappear because most Ecuadorean palms, unless cared for, can regenerate only in forested areas. The large stands of fully grown palms in the pasture lands are made up of the last survivors from former populations that included seedlings, juveniles, sub-adults, and adults; they are the old survivors in damaged populations that have lost their recruitment basis.

As the palm populations are being decimated the species remain protected from extinction only in national parks and reserves. Although the parks and reserves may protect a species from extinction, important genetic variation is lost when the palms disappear from areas outside the protected zones. An example of this is *Jessenia bataua* where it grows at an altitude of 1350 meters above sea level along the road from Loja to Zamora in southern Ecuador. Nowhere else in South America has this species been recorded above 1000 meters. Forest clearance in the area is rapid, and *J. bataua* is likely to disappear in the near future. This will cause loss of genetic variation and possibly prevent its cultivation at such high altitudes in the future.

While it is not the only threat to wild palms, deforestation is the major menace. However, excessive exploitation has also seriously depleted some species. Again the issue is not clear-cut, for in some cases commercialization leads to protection and cultivation of species, while in other cases commercialization leads to destructive exploitation.

Background-information about Ecuadorean palms

Burret's publications (1928–1940) include descriptions of new species and other information about palms in Ecuador. Balslev and Barfod (1987) provided an overview of Ecuadorean palms. The natural history and economic botany of the ivory palms (subfamily Phytelephantoideae) was treated by Barfod (1988). Other publications with information on taxonomy and distribution of palms in Ecuador are: Arguëllo (1984, 1989), Balslev and Henderson (1986, 1987a, 1987b), Balslev *et al.* (1987), Beccari (1916), Blicher-Mathiesen and Balslev (*in press*), Borchsenius and Balslev (1990), Dahlgren (1936), Dodson and Gentry (1978), Glassman (1965, 1970, 1972), Little (1970), Little and Dixon (1969), Moore (1973a, 1980, 1982), Skov and Balslev (1989), and Wessels Boer (1968).

The pollination biology of *Jessenia bataua* in Ecuador was described by García S. (1988) and that of *Phytelephas microcarpa* by Barfod *et al.* (1987).

The economic botany of *Phytelephas aequatorialis* and other palms from the coastal lowland of Ecuador was treated by Acosta-Solís (1944, 1948, 1952, 1961, 1963, 1971). Some of the above publications, and all of the following, include additional information on ethnobotany and economic botany of palms in Ecuador: Alarcón G. (1988), Balslev (1987), Balslev *et al.* (1988), Barfod and Balslev (1988), Barfod *et al.* (1990), Barret (1925), Bianchi *et al.* (1982), Carrión (1970), Cook (1942), Estrella (1988), García E. (1987), García S. (1986), Iglesias (1985), Kvist and Holm-Nielsen (1987), Lescure *et al.* (1987), Marles *et al.* (1988), Orr and Wrisley (1981), Pellizaro (1978), Vickers (1976), Vickers and Plowman (1984), and Wheeler (1970).

The struggles between indigenous people and owners of oil palm plantations in Amazonian Ecuador were described by Anonymous (1986), CONFENIAE (1985), and Jijón (1986). The development of the oil palm plantation industry was treated by Carrión and Cuvi (1985). Three publications deal with fertilization and pests of the African oil palm in Ecuador (Orellana M. 1986, Rivadeneira Z. 1985, and Vera D. and Orellana M. 1986).

Experimental work with the African oil palm (*Elaeis guineensis*) is carried out by Instituto Nacional de Investigaciones Agropecuarias (INIAP) in Ecuador. Henrik Balslev and collaborators at the Botanical Institute, Aarhus University, treat the palms for the Flora of Ecuador. Studies on spread of pests from wild palms to cultivated oil palms are carried out by Palmeras del Ecuador S. A. at Shushufindi. A germplasm plantation with *Bactris gasipaes* has been established by INIAP in Amazonian Ecuador. Information about cultivation and biology of *Cocos nucifera* is presented by MAG (1975).

2. PALMS AND EXTRACTIVISM

Extractivism — the harvest of products from plants that grow in the wild — has been of immense importance throughout human history. It is still widely practiced, especially in the tropics where it contributes significantly to the well being and survival of a great number of people.

Extractivism is carried out not only by people in subsistence cultures. Many wild plants are exploited on a commercial scale, the Pará rubber tree (*Hevea brasiliensis*) being one of the best known examples. As happened with the Pará rubber tree, extractivism often leads to cultivation which then, sometimes, replaces commercial extractivism or, again with the Pará rubber tree as an example, cultivation is carried out alongside extractivism.

Among the palms, wild stands of the African oil palm (*Elaeis guineensis*) in West Africa were exploited for centuries before the Europeans became aware of it. After its discovery the growing European consumption of palm oil was based on exploitation of wild stands of *E. guineensis* for almost one hundred years, but in 1911 large scale cultivation began in Dutch East Indies — the present Indonesia. The palm soon became an important plantation crop in Southeast Asia, especially in Malaysia (Hartley 1977), and today, while Southeast Asia remains its main center, the palm is also cultivated in central Africa, for instance in Nigeria and Ivory Coast, and in some South American countries such as Colombia and Ecuador. Exploitation of wild stands of *E. guineensis* is still practiced in West Africa, where it furnishes fruits, leaves for thatch, and sweet sap from the trunk and inflorescence (Hartley 1977).

Subsistence extractivism
Fruits, fibers, leaves, and other products which are important to indigenous people are often harvested destructively by felling the palms. The resulting depletion of palms may not have been important to nomadic and semi-nomadic tribes that moved around in sparsely populated regions, although it has been suggested that this practice was one of the factors that forced them to move (Anderson 1978). The Waorani indians in Amazonian Ecuador move frequently because of depletion of stands of *Astrocaryum chambira*, the young leaves of which are used to make strings for hammocks *etc.* This may be a recent problem because hammocks are now being exported from the region as souvenirs (Lescure *et al.* 1987). More studies on historic exploitation methods are needed. It is likely that destructive harvest was rare until iron tools were introduced, because even with a *machete* many palms are hard to cut. Today many tribal groups are permanently settled and their territories are restricted, which often leads to excessive exploitation of their natural resources. In the vicinity of Canelos, a Quichua village in Amazonian Ecuador, the once abundant *Jessenia bataua* has disappeared because of over-harvesting of its fruits, and stands of *Ammandra natalia* have been depleted because of destructive harvest for fibers.

In north Brazil, in the territory of the Xirana-teri, a Yanamama tribe who have been settled for 15 years, there has been a depletion of some palm species due to over-use and this tribe now uses few palms (Anderson 1978). The Shipibo tribe of Amazonian Peru have no ritual regulation of palm harvesting and will cut down a species of *Scheelea* to obtain leaves that could have been harvested by climbing. The Shipibo also fell a species of *Oenocarpus* to obtain its fruits, and no use is made of the wood. Bodley and Benson (1979) who studied this tribe concluded that if there was an ethic to protect and conserve forest resources it has not survived and under aboriginal conditions it was probably not required.

Collecting of palm fruits and leaves by felling is not necessary, as other methods of collection of fruits of the cultivated peach palm (*Bactris gasipaes*) have been developed by indigenous people. These methods could be used for wild palms as well. Obviously the only way to obtain some products, for instance palm hearts, starch from the trunk, or timber for construction is to fell the tree.

Although extractivism is often destructive it is not always so. The Siona indians of Amazonian Ecuador extract fibers from *Astrocaryum chambira* leaves by collecting from low palms. The Waorani indians, however, claim that harvest of chambira fibers depletes the palm around their dwellings (Lescure *et al*. 1987). The tagua palm, *Phytelephas aequatorialis*, is carefully conserved by the Cayapa indians of the coastal plain of Ecuador for its many uses (Barret 1925).

Some palms become more common in areas where indigenous groups settle, and some species are often associated with archaeological sites in the Amazon: *Astrocaryum vulgare, Elaeis oleifera, Acrocomia eriocantha, Maximiliana maripa, Orbignya phalerata, Bactris gasipaes,* and *Mauritia flexuosa* (Balée 1988). Of these, *Bactris gasipaes,* and possibly some of the others, are often cultivated. *Orbignya phalerata* is well adapted to disturbance, because its young stages are fire resistant. Some palms may be favored by the nutrient rich black soil, *terra preta,* that is found at archaeological sites as a result of accumulation of human waste. Accumulation of disposed seeds, and the fact that some palms are pioneer species that benefit from clearings made near the dwellings, may also make some of these species more common after human occupation.

Commercial extractivism

Commercial extractivism (*extractive exploitation* of Lleras and Coradin 1988) is common in tropical forests, but few statistics are available to show quantities harvested or economic importance. The registered value of products obtained from six native genera of palms in Brazil amounted to 100 million US$ in 1979 (Table 1).

In Ecuador several wild palm species are exploited commercially; *Prestoea trichoclada* and *Euterpe chaunostachys* supply palm hearts for canning industries in Quito and Borbón; seeds (vegetable ivory) of *Phytelephas aequatorialis* furnish the raw material for button factories in Manta and souvenir workshops in Riobamba and Quito. In Manta, oil is industrially extracted from the seeds of *Attalea colenda,* and *Ammandra natalia* provides fibers that are used in a number of small broom producing industries throughout Ecuador.

Table 1. Products obtained from native palms in Brazil. References: 1. Anderson (1988), 2. Balick (1985), 3. Johnson (1985).

Species	Product	Quantity metric tons	Value 1,000 US $	Year
Astrocaryum aculeatum and *A. vulgare*[2]	Seed oil	8,381	1,800	1980
Astrocaryum tucuma[3]	Seed oil	6,000	-	1982
Syagrus coronata[2]	Seed oil	7,729	1,690	1980
Attalea funifera[3]	Fibers	58,089	-	1982
Mauritia flexuosa[3]	Fibers	1,277	-	1982
Copernicia prunifera[3]	Wax	23,000	-	1982
Euterpe oleracea 3	Fruits (Juice)	84,686	-	1982
Orbignya phalerata[2]	Kernel oil	250,949	71,600	1980
Total sale for six genera[1]			100,000	1979

Several other palm products enter the local markets on a smaller scale. Oil and fruits from *Jessenia bataua*, and fruits from *Ammandra natalia,* are sold on markets in Amazonian Ecuador. Fruits from *Euterpe chaunostachys* are used for beverage and ice cream in San Lorenzo and Borbón. Hammocks and nets of *Astrocaryum chambira* fibers can be found in towns in Amazonian Ecuador and in souvenir shops in Quito. Around Esmeraldas young leaves from *A. standleyanum* are harvested for making hats. Small baskets, made from leaves of *Ceroxylon*, are sold in Andean Ecuador at Easter time.

Iriartea deltoidea is harvested from wild stands near Puerto Quito on the western foothills of the Andes. The trunks are cut into poles which are sold to banana (*Musa x paradisiaca*) plantations on the coastal plain for supporting fruiting plants. The poles made from *I. deltoidea* are said to last 3–5 years when used for this purpose in contrast to those made from other trees which last only 1–2 years. The wood of *Iriartea deltoidea* is also cut and sold to the furniture industry in Guayaquil and Quevedo according to local people from the Puerto Quito area.

These products are exclusively from wild palms, except those from *Phytelephas aequatorialis, Prestoea trichoclada,* and *Ammandra natalia* which are, occasionally, cultivated.

Destructive extractivism

The effects of commercial extractivism on the environment and on the indigenous people vary, but present logging operations in tropical forests are probably the most devastating form of extractivism ever seen, and examples of negative impacts are numerous, also when palms are concerned. *Phytelephas aequatorialis* was the first wild palm species in Ecuador to be exploited on a commercial scale, and export of its seeds started some 130 years ago. Seeds were obtained mainly from wild stands, collected from the ground, or harvested from standing palms, which were often felled to obtain fruits otherwise out of reach. This form of harvesting killed the palm, and fruits so obtained were often to immature to be used. Acosta-Solís (1944) described this harvesting method and strictly opposed it and suggested that it should be forbidden by law and violators punished. In the 1940s it was forbidden to fell *Jessenia bataua* palms in Brazil, and violators were fined and imprisoned (Ranghel G. 1945). Present day exploitation of *Ceroxylon* at Easter time is becoming destructive in certain parts of Ecuador. In the area between Baeza and Cossanga on the eastern slopes of the Andes *Ceroxylon* trees are felled to obtain the young leaves which are used for weaving decorative baskets for Palm Sunday. Felling a *Ceroxylon* palm takes about 15 minutes and yields an average of five leaves which sell at 2000 Sucres (= 3 US$) in Quito.

These examples illustrate what seems all too common; extractivism becomes destructive and depletes the resource that supports it. The palm heart industry provides other examples; in Brazil *Euterpe edulis* has been eliminated in many areas (Nodari and Guerra 1986), and the same is true for another species of *Euterpe* in Costa Rica (Balick 1976). Around Iquitos, in Amazonian Peru, *Mauritia flexuosa* has been depleted following destructive harvest of fruits, of which large amounts are sold on the market (Padoch 1988). The Chilean wine palm, *Jubaea chilensis,* is listed as "endangered" by the International Union for Conservation of Nature (Johnson 1988), partly because it is being felled to obtain sap for wine (Corner 1966, Dugand 1972, Moore 1979).

Two main factors contribute to the destructiveness of some commercial extractivism. First, extractivism is often carried out on land with no recognized ownership, which makes it possible for several companies or individuals to exploit the same resource at the same time, and "take the profit and run" before someone else does it. Second, there is a lack of adequate harvesting tools and methods for which reason palms are often felled instead of harvested by other means to obtain their products.

It appears, that strict control is necessary to secure a sustainable exploitation. Control is difficult in many of the areas where extractivism is practiced and logistic problems and the socio-economic reality means that existing rules are often not enforced. Measurements to reduce "take the profit and run" exploitation should be designed in a way which, as much as possible, reduces the need for control which means that such rules should make it easier and/or more profitable to exploit the resource in a sustainable way than to exploit it in a destructive way. For this purpose the following measures could help:

1. Concessions should be assigned, and only one group of people or one company should be allowed to do large scale commercial exploitation of a certain resource in any given area.
2. Concessions should be long lasting which would make the persons or companies which obtain them motivated for a sustained yield based exploitation.
3. Companies with concessions should pay a certain amount of money (apart from taxes

etc.) every year to be used for re-planting and for compensation to local people if over-exploitation results. Otherwise the money should be paid back to the concession holder at the end of the concession time.

4. Transfer and development of not-destructive harvesting tools and methods should be implemented. These should not only be inexpensive and durable, but they should also provide the easiest way to harvest the product and represent an attractive alternative to felling. For some products, such methods already exist in some areas. In the Amazon estuary fruits from *Euterpe oleracea* are usually harvested by climbing, whereas in Ecuador and Colombia the fruits from the closely related *E. chaunostachys* are commonly harvested by felling.

Nutrient depletion

Most soils in the Amazon are poor in available plant nutrients, but small amounts of nutrients are continuously imported with the rain. One hectare in central Amazonia, with a yearly precipitation of 2,000 mm receives 8.2 kilograms of nitrogen (N), 0.22 kilograms of phosphorous (P), 2 kilograms of potassium (K), and 1.4 kilograms of calcium (Ca) per year (Junk and Furch 1985). While this is a significant contribution in an otherwise nutrient poor area, it is not enough to replace exported nutrients if intensive extractivism of nutrient-rich products is undertaken. Therefore fertilization or long fallow periods are needed. On the other hand nutrient poor products, such as fibers and vegetable ivory, may be harvested with no significant influence on the nutrient reserves in a given area.

Obstacles to extractivism

A major obstacle to more widespread use of extractivism is, that the actual and potential importance of extractivism is often underestimated by development planners. One reason for lack of appreciation is that economic theory has yet to provide models that, in a satisfactory way, describe the long term benefits of leaving the forest in a semi-natural state, rather than clearing it (Leslie 1987, Danida 1988). These long term benefits include products, environmental stability, conservation of species, and social services.

Another obstacle to extractivism is, that in order to obtain ownership to a piece of land in Ecuador a farmer must show that the land is being worked and used for cultivation of crops or cattle raising. A disadvantage of such requirements in real-estate is that forests that could serve other purposes are cleared. The Shuars in Amazonian Ecuador provide an example of the consequences; according to Tseremp Juanka, a shuar who lives in the area, they now raise cattle to keep re-growth off cleared land in order to obtain ownership to the land.

Finally production from wild stands of palms gives unpredictable yields and a heterogeneous quality of the products. These are, seen from the industrialists point of view, obstacles to extractivism. For instance, for palm hearts of *Bactris gasipaes* quality control is favored by technical management of plantations and industrial processing of the palmito (Clement and Mora Urpi 1987).

Effects of extractivism

As shown above extractivism is often destructive, but there are examples of extractivism having led to the protection of a species. One example of this is *Ammandra natalia* in Ecuador which has been depleted in some areas, but at the same time it has been preserved, and its re-growth promoted, in other areas.

Extractivism may lead to domestication of new crop plants, and it may demonstrate the necessity for conservation of the natural vegetation by demonstrating its economic potential. In this way it may secure that more land is being kept in a semi-natural state rather than cleared, for instance in buffer zones around national parks and other protected areas where it may become a useful way to exploit the natural vegetation in a sustainable way. Furthermore, extractivism is often work-intensive and cost-extensive, which is suitable in many tropical countries. Last, but not least, the history and culture of indigenous people is closely related to the natural vegetation in a given area, and extractivism, or cultivation of native plants, are likely to protect their cultural heritage better than an agriculture based on introduced plants.

If extractivism is to become an integrated and sustainable part of future land-use systems in the tropics it needs more attention from researchers and authorities in the concerned countries. More projects to design and start small scale local industries that can process the local products should be undertaken, and research on their influence on the vegetation, nutrient balance *etc.* should be carried out along with it.

3. PALMS IN AGROFORESTRY

Agroforestry is a broad concept for which various definitions have been given. Bene *et al.* (1977) defined it as: "a sustainable management system for land that increases production, combines agriculture crops, tree crops, and forest plants and /or animals simultaneously or sequentially, and applies management practices that are compatible with the cultural patterns of the local populations". This definition includes ideal goals such as increasing overall production and being sustainable and culturally appropriate, which are not always fulfilled in existing agroforestry systems (Raintree 1987). In common use, the term "agroforestry" is applied to any land-use system with a combination of trees, herbs and/ or livestock on the same land and a series of special terms are used to denote different types of agroforestry: "agro-silvi-culture" produces agricultural crops among forest (wild) crops, "silvo-pastoral" systems integrate forest trees and grazing, "agro-silvo-pastoral" systems integrate agricultural crops, forest crops and animal husbandry (Nair 1980).

In the tropics agroforestry systems are diverse and have several advantages compared to cultivation of annual crops. The continuous vegetation cover in agroforestry systems reduces soil erosion, it maintains a more stable micro-climate, it has a more closed nutrient cycling, and trees often grow well on soils that cannot sustain annual crops (Nair 1980, Poulsen 1978). Agroforestry systems resemble natural systems in being more complex than mono-culture systems and they are therefore believed to be less susceptible to pests (Clement 1986). To the farmer a diverse system with many crop species may provide a wide range of daily needs and it makes him less vulnerable to fluctuations in market prices of individual products, and to pests that destroy specific crops.

From the conservationists point of view the advantages of agroforestry are the general environmental services offered by a continuous vegetation cover such as climatic stability, soil conservation, and protection of catchment areas. Sustainable farming systems also produce less pressure on remaining forests, and more species and varieties are protected in a diverse system. With many local crop plants the agroforestry systems act as *in situ* gene banks.

Agroforestry is an old land-use form in many parts the tropics, including Ecuador, but it still needs to be promoted. New settlements in forested areas produce alarming rates of deforestation. The settlers, in Ecuador called *colonos,* often arrive from large cities and have no knowledge at all of agriculture, or they may come from the Andean region where they are used to cultivation methods that are unsuitable for tropical lowlands. These people, as well as the forest, would benefit from programs promoting agroforestry. Such programs could include:

1. Local consultants with good knowledge of the native flora who could advise farmers about potential uses of the flora before forest clearing is initiated.
2. Local nurseries that provide inexpensive seedlings of a range of native species.
3. Demonstration plots.

4. Financial support during early years of farm establishment. Tree crops need years before they produce, and financial aid may be needed to convince farmers to grow them.
5. Enhancement of marketing possibilities for products obtained from forest trees, which would make it more attractive to spare them from felling and incorporate them into the farming systems.
6. Establishment of small local industries to process quickly deteriorating products such as oil fruits and palm heart. This would make it possible to grow a greater variety of crop plants.

Agroforestry systems resemble natural systems in some ways and have advantages compared to such land-use systems as pastures or mono-crop agriculture. But even if agroforestry can help reduce the problems of deforestation it cannot solve them. Agroforestry systems that replace natural forests still reduce biological diversity but agroforestry systems can also be established on land which has already been degraded.

The International Council for Research in Agroforestry (ICRAF) in Kenya is the main international center for research in this field but many national institutions throughout the tropics, such as Instituto Nacional de Investigaciones Agro Pecuarias (INIAP) in Ecuador, also carry out research on agroforestry. Some results obtained by the Payamino station of INIAP near Coca in Amazonian Ecuador are presented by Bishop (1980) and Estrada *et al.* (1988). The journal *Agroforestry Systems* provides valuable information on the subject.

The palm component in agroforestry systems
Palms are suitable for agroforestry and integration of palms into the agricultural systems may proceed faster and easier than introduction of new and unknown species. In lowland Ecuador several factors make palms particularly suitable for agroforestry:
1. Palms are already there.
2. Among indigenous people, as well as *colonos,* palms enjoy a special status because of their usefulness and ornamental value.
3. Many palms can grow on land unsuited for other crops, *e.g.* acidic swamps.
4. Many palm products, such as fibers and vegetable ivory, can be stored for a long time. This is important to small farmers because they need to store their products until transportation is available and sufficient quantities have been gathered to justify the cost involved in transporting the products to the market.
5. Many palms are multipurpose trees (Johnson 1983) which, for instance, at the same time produce fruits for consumption and stabilize the soil through an extensive root system (Burley and Carlowitch 1984).
6. Palms produce less shade than dicotyledons (Johnson 1983) and therefore permit more crops to be grown underneath them.
7. The stature of palms permits a greater planting density than is feasible with spreading dicotyledons (Johnson 1983).
8. Palms are less vulnerable to fire damage than conifers and dicotyledons, because of a different arrangement of their vascular tissue. This is an advantage if fire is used to

control weeds (Johnson 1983). Resistance to fire means that palms, whether it is intended or not, often become part of land-use systems that replace forest.

9. Some palms reproduce quickly and easily from basal suckers (Johnson 1983).

10. Palms may play an important role in recovering deep soil nutrients (Anderson 1988), however, many palms have a dense but mainly superficial root system (Corner 1966, Granville 1974 , Mora Urpi *et al.* 1982). Dense superficial root systems stabilize soils and trap nutrients and they reduce leaching, but they are not able to absorb nutrients from deep soil layers and may compete seriously with annuals. New adventitious roots are continuously formed while old roots die (Corner 1966) which may increase the microbial activity in the soil by adding organic material below the surface and increase soil aeration.

Rooting patterns must be considered when designing agroforestry systems. The optimal situation is a continuous root mat with little overlap within the same horizons which reduces leaching and competition.

These points about palms in agroforestry may apply to many palms but it is difficult to generalize about the role of individual species in an agroforestry system. In the wild palms occur in a variety of habitats and niches; some are canopy species in closed forest (*Jessenia bataua*) and others occur in the sub-canopy (*Ammandra natalia*), some grow on acidic, nutrient poor, water logged soils (*Mauritia flexuosa*) and others on regularly inundated, nutrient rich soils (*Euterpe chaunostachys*), and finally some are pioneer species adapted to colonize seasonally inundated flood plains (*Astrocaryum jauari*). The habitat, and the role of a palm species within the habitat, offers suggestions to where it can be grown, and how it should be incorporated into an agroforestry system, but, because the natural distribution of palm species is influenced by competition between species, it can not be deduced that, *e.g.* a palm that is found only in swamps can not be cultivated on *terra firme*. Likewise some sub-canopy species may prosper much better if exposed to full sunlight.

Theory can help when designing agroforestry systems, but so far experimental cultivation is the only reliable method to establish a species composition that will result in positive interaction (Nair 1980). Experimental cultivation has sometimes revealed unexpected results. For example, the yield of coconuts (*Cocos nucifera*) increased when it was inter-planted with cacao (*Theobroma cacao*) (Nair 1979).

Few palm species are grown in agroforestry systems in Ecuador. The most commonly used one is the peach palm, *Bactris gasipaes*. Other examples are *Phytelephas aequatorialis*, *Ammandra natalia*, *Cocos nucifera,* and *Elaeis guineensis*. The latter is essentially a mono-crop in Ecuador, but in its establishment phase other plants are sometimes inter-cropped. Other species of palms, which are left when forest is cleared, automatically become elements in various land-use systems which replace natural forests.

The peach palm — *Bactris gasipaes*

The peach palm, *Bactris gasipaes,* domesticated by indigenous people in South America, is now widespread in the Neotropics and found only in cultivation. In Ecuador *B. gasipaes* is grown throughout the humid lowlands up to about 1000 meters above sea level, most commonly in Amazonian Ecuador and less frequently on the coastal plain. It is cultivated mainly for its fruits, which are cooked and eaten, or used for making a beverage known as *chicha*. The fruits of *B. gasipaes* are sold, raw or prepared, on the markets in the lowland but commercialization has not reached the same dimensions as in Costa Rica and Colombia. There, plantations and small farmers supply palm hearts and fruits for canning for domestic and foreign consumption (Clement and Mora Urpi 1987). *Bactris gasipaes* is grown in a variety of associations, for example as shade tree in coffee (*Coffea arabica*) and cacao (*Theobroma cacao*) plantations and together with bread fruits (*Artocarpus altilis*) and citrus (*Citrus* spp.). The agroforestry system of the Siona-Secoya indians in Amazonian Ecuador is a multi-strata garden that is established after clearing and sometimes burning. Sweet potatoes (*Ipomoea batatas*), pineapples (*Ananas comosus*), and taya (*Xanthosoma* sp.) form the ground layer. Cassava (*Manihot esculenta*), sugarcane (*Saccharum officinarum*), and maize (*Zea mays*) followed by papaya (*Carica papaya*), form the middle layer. The top layer is formed by *Bactris gasipaes* and species of the leguminous tree *Inga*. Many crop species are used in indian gardens: in the village of Shushufindi 54 species and 112 varieties are grown for food (Vickers 1978).

Experimental work with *Bactris gasipaes* as a component in agroforestry systems is being carried out several places in Brazil, for instance at the Instituto Nacional de Pesquisas da Amazonia (INPA) in Manaus and at the Cacao Research Center (CEPEC) of the Brazilian Cacao Commission CEPLAC. The experiments at CEPEC concerning the use of *B. gasipaes* as a shade tree in cacao plantations has been very successful, and the palm may now be recommended for this use throughout the Brazilian Amazon (Clement 1986).

At the INIAP field station near Coca in Amazonian Ecuador a germplasm plantation of *Bactris gasipaes* has been established from seeds collected in Colombia, Peru, and Ecuador. According to V. Chalá and R. Mera, of the INIAP station, the aim of the germplasm plantation is, apart from being a gene bank, to develop spineless varieties (of which some already exist), pest resistance, adaptation to acid soils, tolerance to climatic fluctuations, and higher yield of fruits and palm heart.

In Ecuador *Bactris gasipaes* is widely accepted by indigenous people and *colonos* and furnishes many products such as palm hearts, fruits, and durable wood. The palm hearts could become a valuable export crop, and the fruits can be used for direct human consumption, for flour production, for animal ration, and for oil production (Clement and Mora Urpi 1987). This, combined with its high yields and the beneficial associations with cacao, suggests that the peach palm will gain importance as an agroforestry and plantation crop in the future.

Figure 1. a. A silvo-pastoral system on the west-Andean foothills along the highway from Quito to Sto. Domingo. The tagua palm, *Phytelephas aequatorialis,* is the woody component, and cattle and pasture is the pastoral component. **b.** A mill used to grind leftovers from *Phytelephas* seeds at the button factory in Manta on the coastal plain of Ecuador.

The tagua palm — *Phytelephas aequatorialis*
The tagua palm, *Phytelaphas aequatorialis,* is widespread on the Ecuadorean coastal plain up to 1500 meters above sea level, where its seeds, that give the "vegetable ivory", are commonly collected from natural stands. *Phytelephas* palms often occur scattered in pastures as the woody component in a simple silvo-pastoral system and occasionally it is cultivated in agroforestry systems, for instance in association with maize (*Zea mays*), citrus fruits (*Citrus* spp.), coffee (*Coffea arabica*), banana (*Musa x paradisiaca*), sugarcane (*Saccharum officinarum*), papaya (*Carica papaya*), and the calabash tree (*Crescentia cujete*) (Fig. 1a). The tagua palm is a suitable crop for small farmers, to whom it provides a cash crop that can be stored and sold at any time of the year.

Up until the Second World War there was a sizable export of vegetable ivory from Ecuador for the production of shirt buttons in USA and western Europe. After the introduction of synthetic materials for this production the world market for vegetable ivory declined, but now there is a growing demand for buttons made from this natural product in Italy, USA, Japan, and West Germany. If this trend continues, a lack of supplies of *Phytelephas* seeds can be foreseen, and increased cultivation may result. At a button factory in Manta on the coastal plain of Ecuador, all residues obtained when the buttons were cut from the seeds were ground in a small mill and the resulting flour was sold to be mixed with cotton press cake and used as fodder for cattle (Fig. 1b). Seeds of *Phytelephas* contain 70% mannan-polysaccharides, 7.5% cellulose, and 22.5% undetermined material (Timell 1957) and the flour is therefore mainly a source of calories. The same use could be made for other palm seeds, for instance those of *Mauritia flexuosa* and *Ammandra natalia.*

The fiber palm — *Ammandra natalia*
Ammandra natalia supplies the fibers for most brooms in Ecuador. Fibers are mainly harvested from wild palms, but in the province of Morona-Santiago in southern Amazonian Ecuador *Ammandra natalia* is frequently found in agroforestry systems in association with peach palm *(Bactris gasipaes),* cassava *(Manihot esculenta),* taya *(Xanthosoma* sp.), cacao *(Theobroma cacao),* coffee *(Coffea arabica),* papaya *(Carica papaya),* sugarcane *(Saccharum officinarum),* citrus fruits *(Citrus sp.),* maize *(Zea mays),* sapote *(Matisia cordata),* and achiote *(Bixa orellana).* It grows well, both in shade and fully exposed sun, and may occupy different niches in an agroforestry system, depending on its height. It is often the only woody component in pastures. This species is treated further in Chapter 10.

4. PALMS IN PLANTATIONS

Plantations are mono-cultures of trees. Compared to a herbaceous ground cover, a continuous tree cover has a number of advantages as discussed in the chapter on palms in agroforestry. Plantations offer advantages also when compared to extractivism and agro-forestry, especially if the product, *e.g.* oil-containing fruits, deteriorate quickly. If a plantation is large enough it may produce enough fruits to make the establishment of an oil mill and the needed infrastructure economically feasible. In addition plantations usually have predictable yields and homogeneous products. On the negative side planta-tions are often associated with ecological and social problems. Mono-cultures that replace natural forests result in loss of biological diversity, and the intense cultivation of a single plant species rapidly depletes the nutrient reserves of the soil, pest problems can be serious and result in extensive use of pesticides. For the African Oil palm, *Elaeis guineensis,* it was predicted, and it has now been confirmed, that plantations in South America would become heavily infested, because many native palm species could supply pests (Hartley 1977). *Elaeis guineensis* was more or less free of pests until the Second World War, but as the planted area grows pest problems increase allover the tropics.

In Amazonian Ecuador social problems related to *E. guineensis*plantations are growing. Plantations were established on land occupied by indigenous people, which caused many protests by these people, who fear that pollution from pesticides and oil mills will destroy their hunting and fishing grounds (CONFENIAE 1985).

Apart from *Elaeis guineensis* the only palms which are cultivated on a commercial scale in Ecuador are hybrids between *E. guineensis* and the American oil palm *E. oleifera,* and to a limited extent *Cocos nucifera.* Experimental plantations have been established with *Bactris gasipaes* and *Elaeis oleifera* by INIAP (Carrión and Cuvi 1985).

The African oil palm — *Elaeis guineensis*
To satisfy the national demand for vegetable oil, and to obtain a new export product, the Ecuadorean government promoted cultivation of *E. guineensis* (Fig. 2) from the early 1960s through favorable credits and by establishing an oil palm research program. The research was primarily carried out by INIAP on its experimental station near Sto. Domingo on the coastal plain where the first plantation of *E. guineensis* was established in 1953/54 along the road from Sto. Domingo to Quinindé. Until 1960 that plantation remained the only one in Ecuador.

During the 1960s the area planted with oil palms increased slowly, but in the 1970s the growth accelerated (Carrión and Cuvi 1985). Until 1978 all plantations were on the coastal plain, but in 1979 plantations were started in Napo province in Amazonian Ecuador. The development of oil palm plantations in Amazonian Ecuador is dominated by two large companies, whereas the industry on the coastal plain is made up of a large number of small plantations. In 1981 there were 230 oil palm plantations each holding less than 300 hectares and 15 palm plantations holding more than 300 hectares (Carrión and Cuvi 1985).

In Ecuador oil palms are grown mainly in mono-cultures though usually in association with a legume cover-crop (*Pueraria* sp.) that protects the soil against erosion and provides nitrogen. Often, other crops are cultivated between the rows of palms in newly established plantations. Soya bean (*Glycine max*), banana (*Musa x paradisiaca*), papaya (*Carica papaya*), and pineapples (*Ananas comosus*) are grown to provide the farmer with a cash income until the palms start to produce. This method is usually referred to as *inter-cropping* or, since it is only carried out during the early years after planting, *establishment inter-cropping* (Hartley 1977).

In West Africa inter-cropping is common and trials there showed that establishment inter-cropping with cassava (*Manihot esculenta*), maize (*Zea mays*), upland rice (*Oryza* sp.) and bananas (*Musa x paradisiaca*) for 1–3 years usually increase the yield of the palms. The trials were, however, carried out on recently cleared forest land, which probably hold larger nutrient reserves than areas under cultivation (Hartley 1977).

In Ecuador African oil palm plantations are often used for cattle grazing but this should be avoided in young plantations if each palm is not carefully protected because the cattle's hoofs may damage the roots (Hartley 1977). The superficial root system of old palms may also be damaged in this way, which, according to Hans Andersen who is a farmer in the Sto. Domingo area,, reduces nutrient uptake and increase the incidence infections.

Oil palm plantations in Ecuador do, in some cases and during some periods, serve other purposes than production of fruits but the possibilities offered by inter-cropping appear to be under-exploited. More research on favorable combinations, and more use of the research that has already been done, for instance regarding the apparently beneficial combination of cacao and oil palms (Hartley 1977), are needed.

The most important products from the African oil palm are the mesocarp oil, which is used for human consumption as cooking oil and fat, and the endosperm (kernel) oil, which, at least in part, is used in the cosmetic industry. Minor products from this palm include; the endocarp which is hard and bony and sometimes used as fuel in the oil mills or as pavement on plantation roads, the mesocarp presscake which is used for fuel, the endosperm presscake which is occasionally used as forage for cattle and chickens, and the empty infructescences which are burned and the ash used as fertilizer (Carrión and Cuvi 1985). When the palms grow too high to be easily harvested, or too old to give satisfactory yields, they are felled and replaced. We have no indication that the felled palms are used although exploitation of them for sweet sap (palm wine) and palm hearts could possibly be profitable.

The American oil palm — *Elaeis oleifera*

The American oil palm *Elaeis oleifera*, distributed in Central America and northern South America, was not reported from the wild in Ecuador until recently when some stands of it were discovered in the Amazonian lowlands near Taisha (Balslev and Henderson 1986). Two decades earlier, however, in 1965 a 39 hectare experimental

Figure 2. Fruits of the African oil palm, *Elaeis guineensis,* which was introduced to Ecuador in 1954 and now is the most important cultivated palm in the country.

plantation of *E. oleifera*, apparently based on imported seeds, was established by the
Sto. Domingo INIAP-station to provide material for cross breeding with *E. guineensis*.
Six hundred hectares have now been planted with the resulting hybrid (Carrión and Cuvi
1985). These hybrids seem to have several advantages compared with *E. guineensis*;
their oil is richer in unsaturated fatty acids, their annual height increment is lower, and
they are resistant, or somewhat resistant, to several pests of *E. guineensis*, and their
yield appears to be more or less equal to that of *E. guineensis* (Meunier and Hardon
1976).

The coconut palm — *Cocos nucifera*

The coconut palm, *Cocos nucifera,* is used in many different land-use systems in
contrast to *Elaeis guineensis* which is found only in plantations or occasionally as an
ornamental. In Ecuador the coconut palm is cultivated mainly on the coastal plain, though
it is now becoming more frequent in the Amazon, where it is commonly grown by small
farmers. There are coconut mono-culture plantations on the coastal plain of Ecuador, for
example along the road from Guayaquil to Jipijapa, but more often *Cocos nucifera* is
found mixed with a variety of other perennials and annuals in agroforestry systems, or as
an ornamental. It is planted in rows on the dikes that separate rice fields along the road
from Guayaquil to Naranjal. Common names are *Coco* and *Cocotero* (Balslev and
Barfod 1987).

The main product of the coconut palm is its fruits, which are sold on every market-
place throughout Ecuador for its liquid endosperm, the coco milk, which is drunk, and
the hard part of the endosperm which is eaten, used in ice cream, for sweet coco pasta,
and for dried coco flour. Apart from this the coconut has several minor uses: in many
places household articles, trinkets, and souvenirs are made from the hard, bony endocarp
and sold; the Cayapas indians in western Ecuador use immature, possibly aborted, fruits
of *C. nucifera* as a sterilizing remedy, crushing, boiling, and drinking a decoction of the
husk twice daily during menstruation to stop the periods (Kvist and Holm-Nielsen 1987);
on the market-place in Otavalo the coconut husk is sold as a remedy against abdominal
pains and small brooms are made from the leaf-base fibers; on Isla Puná *Cocos nucifera*
is used as a remedy against intestinal parasites and some kidney disease and its leaves are
used for thatch (Balslev *et al.* 1988).

But in general *Cocos nucifera* is used far less in tropical America than in Southeast
Asia, where it is the most important crop plant in many areas. Exploitation of the husk
fibers from the mesocarp of the fruits is an important industry in Asia. In the beginning
of the 1970s 295,000 metric tons of fibers were harvested annually, and used for manu-
facturing ropes, carpets, sacks, *etc.* (Grimwood 1975). More recent inventions include
the use of trunks of old palms for roof tiles, laminated timber, wood block floors, *etc.*
(FAO 1986). The use of coconut husk fibers, the extraction of oil from the endosperm,
and the tapping of sap from the inflorescences have not been recorded in Ecuador. When
comparing the uses of *C. nucifera* in Ecuador with those in Asia, where such practices
are common, it is evident that the palm represents an under-exploited resource in Ecuador.

5. PALMS AS ORNAMENTALS

Throughout Ecuador a variety of palms are cultivated as ornamental street trees (Table 2). They are grown on almost every plaza and in many gardens and the ornamental value of palms is probably one of the main reasons why palms are often left standing when forest is cleared for agriculture. Many ornamental palms serve other purposes as well, and a clear distinction between ornamentals and crop plants cannot always be made. One example is *Parajubaea cocoides* which is known only in cultivation (Moraes and Henderson 1990); it is one of the most appreciated ornamental street trees in the towns of the Andean region of Ecuador, but it is also grown for its fruits which are edible and known as *cocito,* and trinkets are carved from the hard endocarp of its fruits.

Table 2. The palms most commonly grown as street trees and garden ornamentals in Ecuador.

Species	Street trees	Garden palms	Origin
Aiphanes caryotaefolia		x	introduced from Colombia
Ammandra natalia		x	native
Bactris gasipaes		x	native
Bentinckia nicobarica	x		introduced from Nicobar Islands
Ceroxylon spp.	x	x	native
Chamaedorea poeppigiana		x	native
Chamaerops humilis		x	introduced from Europe
Chrysalidocarpus lutescens	x	x	introduced from Madagascar
Cocos nucifera	x	x	native ?
Elaeis guineensis	x	x	introduced from Africa
Jubaea chilensis	x	x	introduced from Chile
Livistona chinensis	x		introduced from Asia
Mauritia flexuosa	x	x	native
Parajubaea cocoides	x	x	native
Phoenix canariensis	x	x	introduced from Canary Islands
Phoenix reclinata	x		introduced from Africa
Pritchardia pacifica	x		introduced from the Pacific
Roystonea regia	x		introduced from West Indies
Sabal palmetto		x	introduced from North America
Trachycarpus fortunei	x	x	introduced from Asia
Washingtonia robusta	x		introduced from North America

Figure 3. Climbing of *Bactris gasipaes* with wooden triangles attached to the trunk with ropes. The triangles may be raized by putting the weight of the climber on one of the triangles and raizing the other. Photograph taken at Manaus, Brazil, by Ulla Blicher-Mathiesen.

6. HARVEST METHODS FOR PALMS

Palms cannot be pruned to remain low and their harvest is therefore increasingly difficult with age, and the question of how to harvest palm fruits and other palm products is therefore of great importance. Efficient methods that use inexpensive and durable equipment are needed in order to stop the widespread practice of felling wild palms. Development of harvest technology and -methods for palm plantations is also important. This is demonstrated by, for instance, *Elaeis guineensis* which must be replaced with young palms when it grows to high because there are no methods to harvest the fruits of very tall palms. Breeding programs may help reduce the problem by providing hybrids or varieties with slow growth such as the *E. guineensis* x *oleifera* hybrid and dwarf varieties of *Cocos nucifera*.

A number of harvest methods exist, some of which are too troublesome whereas others deserve to be more widely known, but because species of palms differ in many ways, the same harvest method cannot be used for them all (Fig. 3). When *Elaeis guineensis* is grown in plantations, its fruits are usually harvested from the ground by cutting the peduncle, in young trees using a metal chisel with a wooden handle, in old and tall trees using a curved knife mounted on a long pole. *Jessenia bataua* has a very strong and fibrous peduncle and cannot be harvested from the ground because the knives available at present are not strong enough to cut through the peduncle. When a palm is harvested from the ground the fruits will fall to the ground which may produce injuries and lead to faster deterioration.

Most harvest methods involve climbing and, once climbed, many palms can be harvested with a *machete,* and the fruit bunch can then be lowered to the ground with a rope. Climbing may be done with the use of ladders or with steps tied to the trunk. Steps on coconut palms may be made from the husk of a fruit (Boldt 1982) or it can be "ingeniously made from partly split leaf-stalks, the split fibrous stalk making the rope which attaches the unsplit leaf-base as the step" (Corner 1966). Occasionally steps are cut into the trunks of coconut palms, a practice that may easily lead to infections of the trunk.

Ropes are used in a number of different ways. The most simple one is a loop in which the climbers feet are placed and the feet and hands are then moved alternately, during which the loop helps to press the feet close to the trunk while mowing the hands. Instead of a rope a twisted palm leaf may be used (Strudwick and Sobel 1988). Another method, which is not suited for smooth trunks, is to make a larger loop and place it around the trunk and the shoulders, and then, in turns, move the rope and the feet. A third method employs two ropes; one for one foot and the other for the thigh of the opposite leg (Hartley 1977, Corner 1966).

Bactris gasipaes has a spiny trunk and can only be climbed using such methods that keep the climber at some distance from the trunk. A simple method is used by the Waorani indians in Amazonian Ecuador who, when they plant a *B. gasipaes*, also plant a *Cecropia* tree next to it, which they then climb instead of the palm. A more laborious way is a scaffold built around the trunk. As the palm grows, the height of the scaffold can be increased by adding more sections and the fruits can be harvested from it (Barbosa Rodrigues 1903a).

An ingenious device used for climbing spiny palms uses two independent triangles on which a piece of rope is used to make a loop around the trunk and the top part. When the basal part of the triangle is lowered the rope will become tight and hold the triangle in an angel away from the trunk. The triangles are moved one by one, the upper one while standing on the lower, and the lower one while sitting on the upper. This method for climbing was observed by our colleague, Ulla Blicher-Mathiesen, in Brazil. In South India, at the J. B. S. Haldane Research Center, work is carried out to develop a so-called "Palm Tree Climber" which is a new portable instrument that may eliminate much of the hard work and pure danger involved in harvesting palms (Davis 1984).

Once a harvester has climbed a trunk, he may proceed to other palms without climbing up and down. In Sri Lanka ropes are strung out between coconut palms and the harvester then walks on these ropes from palm to palm (Corner 1966). Arecanut palms (*Areca catechu*) are often grown in dense stands in plantations or agroforestry systems in Asia and their harvesting is done by climbing one palm and then swinging from palm to palm, the next palm being reached by pulling it close with a hook. More than one hundred palms may be harvested without the harvester touching the ground (Bavappa *et al.* 1982).

Two climbing devices, often used by biologists and foresters, are spurs and swiss tree grippers. Spurs, which are mounted on the feet, are steel rods bent into a semicircular shape and with spines welded on the inner side. When the climber puts his weight on the foot and the spur the spines will be pressed into the trunk and hold the climber. A security belt is kept around the trunk and fastened to the chest which secures the climber from falling backwards. The spines of the spurs can damage the trunks and cause infections with, for instance, larvae of the beetle *Rhynchophorus palmarum*, but there are no published records of this.

Swiss tree grippers are made of two separate steel braces, mounted with holders for the feet. The adjustable steel braces are mounted around the trunk, and when moving them one by one, it is possible to climb the palm. A security belt prevents the climber from falling backwards.

Apart from biologists and foresters, monkeys may be trained to harvest fruits from palms. This practice has been used in Southeast Asia for harvesting fruits from arecanut and coconut palms (Bavappa *et al.* 1982, Corner 1966).

7. ASTROCARYUM JAUARI

Astrocaryum jauari is a conspicuous element in the vegetation of seasonally inundated areas along many rivers in the Amazon but very little has been published about it. Locally its fruits are used as fish bait, and, for a period, *A. jauari* was the basis for a small oil industry on the Rio Negro in Brazil. Today natural stands of it along the Rio Negro are exploited for palm heart. *Astrocaryum jauari* does not play a major role in the everyday life of indigenous people in Ecuador or in the cash economy, but indirectly it is of great importance because it provides thousands of tons of fruits which are consumed by various species of fish in the area which, in turn, provide protein to the people who live along the rivers. Because of this intricate role in the ecosystem, *Astrocaryum jauari* fruits could prove useful in supplying fodder for fish farming.

Taxonomy
The genus *Astrocaryum* belongs to the palm subfamily Arecoideae of which all members have pinnate leaves, are monoecious, and have inflorescences with flowers in groups of three, two staminate and one pistillate, or evidence of such groups (Uhl and Dransfield 1987). *Astrocaryum* comprises 47 species of which four species are known from Ecuador; *Astrocaryum chambira, A. jauari, A. murumuru,* and *A. standleyanum* (Balslev and Barfod 1987). In Ecuador, *Astrocaryum jauari* is found only in the Amazonian part and it can be distinguished from the other Ecuadorean *Astrocaryum* species by its glabrous fruits. The other species have fruits with either a mealy (*A. chambira* and *A. standleyanum*) or spiny (*A. murumuru*) epicarp. *Astrocaryum jauari* was described by Martius (1823), but no type specimen was designated.

Vernacular names
Oco-be-to and *Huiririma,* both Siona indian names, are the only know vernacular names for *A. jauari* in Ecuador (Balslev and Barfod 1987). The following list includes names from other areas. After each name follows, in parentheses, the area where the name is used or the language in which it is used and the literature reference from where the information was obtained. Names in bold are the ones most commonly used.

Awarra (Brazil, Guianas, Civrieux 1957); *Coqueiro Javari* (Brazil, Pinheiro and Balick 1987); *Corozo* (Venezuela, Civrieux 1957); *Diabexta* (Tucano, Dahlgren 1936); *Guara* (Colombia, Glassmann 1972); *Huiririma* (Siona, Ecuador, Balslev and Barfod 1987; Peru, Pinheiro and Balick 1987, MacBride 1960); *Jamari* (Brazil, Pinheiro and Balick 1987); **Jauari** (Brazil, Pesce 1985); *Javarí* (Amazon region, Schultes 1977); *Liba-awarra* (Surinam, Wessels Boer 1965); *Macanilla* (Venezuela, Markley 1949); *Mauizi* (Wayumará-carib, Dahlgren 1936); *Oco-be-to* (Siona, Ecuador, Balslev and Barfod 1987); *Palmeira Jauari* (Brazil, Pinheiro and Balick 1987); *Rojti* (Kaiapó, Dahlgren 1936); *Sauari* (Macusi, Guiana, Dahlgren 1936); *Sawarai* (Guiana, Dahlgren 1936); *Soela-awarra* (Surinam, Wessels Boer 1965); *Yahuarhuanqui* (Shipibo, Peru, Bodley and Benson 1979); *Yauaranga* (Peru, Pinheiro and Balick 1987); *Yauari, Yavari, Yahuarí* (Colombia, Dugand 1940).

Figure 4. *Astrocaryum jauari* growing on the margin of a blackwater lake in the Cuyabeno Reserve in Amazonian Ecuador. In the foreground there are small individuals of the palm. Note the fuzzy appearance of the leaves and the slender trunks that may reach 20 meters height.

Morphology

Astrocaryum jauari is a medium- to large sized palm with a trunk up to 20 meters tall, and up to 30 centimeters dbh (Fig. 4). We have observed solitary individuals only, but some authors and collectors mention that it may be multi-stemmed (Wessels Boer 1965; *Henderson and Nascimento de Lima no. 605*). Its internodes are heavily armed with spines that point in every direction but older trunks may be spineless on the lower part. The leaves are pinnate, about five meters long, erect to spreading, and number 5–15; the pinnae, numbering 100–150 per side, point in various directions and give the leaf a fussy appearance; leaf sheaths, petioles, rachis, and pinnae margins have spines. The inflorescences are monoecious, racemose and interfoliar, erect, and up to 2.5 meters long. From 2–5 inflorescences may be found on a single palm but in Ecuador three inflorescences seems most common, although Piedade (1985) found an average of four in the lower Rio Negro area. The fruits are green to yellow, glabrous, ovate, and up to four centimeters long; the single seed is enclosed in a hard endocarp and a thin and rather dry meso- and pericarp. The root system in juvenile palms, up to 5–6 years of age, is shallow with the roots not penetrating below 50 cm but adult palms have roots that reach more than one meter into the ground (Schlüter 1989). Pneumatophores or other morphological adaptations to inundation have not been observed but Schlüter found aerenchymatic tissue in the roots and observed both aerobic and anaerobic metabolism in seedlings. Adults were not studied.

Distribution

Astrocaryum jauari is widespread in the Amazon basin and is also found along the Río Orinoco and its tributaries and along rivers in the Guianas. It grows only in seasonally inundated areas in the lowlands, the highest known record being from Ecuador at 230 meters above sea level, and the lowest one near sea level at Rio Tocantins in eastern Brazil (Fig. 5). Rainfall patterns in its distribution area vary from per-humid in Ecuador to seasonal with a dry season in various parts of Brazil with mean annual precipitation in its distribution area varying from 1523–2623 mm, and mean annual temperature from 24.8–27.2°C. Climatic diagrams A, B, C, D, F and 5 in Chapter 12 show data from localities close to where *A. jauari* grows.

Habitat

Astrocaryum jauari occurs in areas which are flooded by blackwater rivers such as the Rio Negro, whitewater as Rio Solimões-Amazonas and Rio Purus, and clearwater as Rio Tocantins and Rio Tapajos. Blackwater is tea colored, poor in nutrients, and has a low pH (3.2 in Rio Negro); whitewater is rich in sediments and nutrients, and have a pH of about 7; clearwater is intermediate and rather poor in nutrients and sediments (Goulding 1980, Junk and Furch 1985). Areas flooded by blackwater- and clearwater rivers, the so-called *igapó,* are infertile, often sandy, and areas flooded by whitewater rivers, the so-called *várzea,* are fertile, often clayey. Thus growth conditions for *A. jauari* vary considerably in terms of soil types, acidity, and available nutrients.

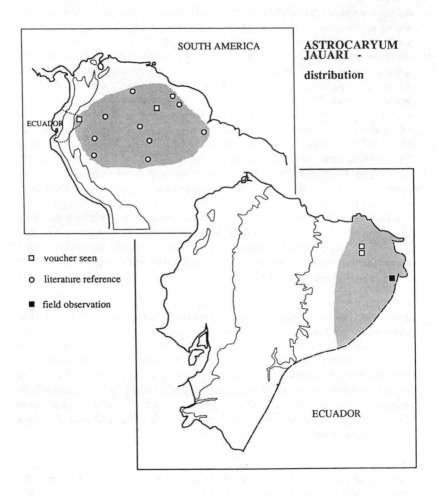

Figure 5. Distribution of *Astrocaryum jauari* in South America and in Ecuador. Its total distribution covers the Amazon basin. In Ecuador it is restricted to the lowlands east of the Andes. The sign for 'vouchers seen' shows the collecting localities for the herbarium specimens cited at the end of this chapter. The sign for 'literature reference' show sites which were given in the literature with such exactitude that they could be placed on the map. The sign for 'field observations' shows such localities where we have observed, but not collected the palm.

In Ecuador we have observed *A. jauari* only along the margins of blackwater lakes and - rivers. In the lower Rio Negro, Piedade (1985) found *A. jauari* in areas which are flooded from 30–340 days per year. The high H_2S levels found in *várzea* water can cause necrosis in the leaves of juvenile plants and Schlüter (1989) found more than 50% mortality in juveniles that had been submerged for more than 200 days under such conditions, whereas no obvious damage was observed in plants that had been flooded for less than 200 days. In *igapó* areas, where H_2S levels are low, juveniles appeared healthy, even after 300 days of flooding.

Apparently *Astrocaryum jauari* can compete successfully only in seasonally flooded forests, where dramatic changes in the environment occurs during the year, from periods with several meters of inundation to periods of drought. Even short dry spells may seriously influence the vegetation in these areas because some of the soils found there dry out quickly and the conditions on a sandy *igapó* may be desert like during dry periods (Junk 1989, Prance 1979).

In the Amazon estuary *Mauritia flexuosa* and *Euterpe oleracea* palms are part of the vegetation succession on the mud-banks which are deposited by the river. Above the Rio Xingu *A. jauari* replaces these two palm species as a pioneer species on the new mud-flats on the river margins, and *Mauritia flexuosa* then grows further away from the rivers (Moore 1973b). *Mauritia flexuosa* and *Euterpe oleracea* are not adapted to occasional drying of the soils and because they rely on pneumatophores for oxygen supply to the roots they may not tolerate the long periods of continuous flooding found upriver. Under such conditions *A. jauari* competes successfully and is part of such vegetation successions as those described by Huber (1906) from the Rio Purus where the river deposits large amounts of silt. The first trees to appear are species of *Cecropia* and in the shade of these, a number of other tree species, including *A. jauari*, germinate. Later *A. jauari* becomes more important and forms a mixed transition forest with a canopy about 12 meters high which may develop into a *Jauari*-forest — a high mixed forest with *A. jauari* as one of the dominant species. The *Jauari*-forest has slender, up to 30 meters tall, semi-deciduous trees with light foliage and it is inundated every year, often with several meters deep water. Deposition of large amounts of silt may kill the *Jauari*-forest and the succession will then start over again with a *Cecropia* forest, but if silt is added more slowly the succession may continue, and the forest then becomes gradually less influenced by the inundations as the land builds up. In the resulting forest, though containing many species from the *Jauari*-forest no, or only old, individuals of *A. jauari* grow. Most of the dominant species in the *Jauari*-forest, along the lower parts of the Rio Purus are also found along Solimões - Amazonas and all the way down river to the Rio Xingu.

Large *igapó* areas in the upper Rio Negro region are dominated by *A. jauari*. Several factors, alone or in combination, could explain this; abundance of dispersers, absence of seed predators, or fire (Goulding 1983). Fire occurs occasionally in the upper Rio Negro region and it is likely that *A. jauari,* like many other palms, is somewhat fire resistant, but it is doubtful whether it can sprout again after fire.

Population structure

In an *igapó* area at Laguna Cuyabeno in Amazonian Ecuador we made a 0.1 hectare (20 x 50 m) plot to measure densities and population structure in one *A. jauari* stand. The stand had a fairly open and diffuse canopy and covered a narrow band, about 20 meters wide, that marks the outer edge of the continuous tree-vegetation towards the seasonal lake. Within the study plot the ground was covered with a thick layer of old *A. jauari* leaves and apart from *A. jauari* seedlings and saplings there was almost no undergrowth. Similar conditions were observed in other stands and we suspect that the exclusion of other species may be due to chemical compounds released by the palms or the decaying leaves. The soil was clayey and the pH value in the water was 5.3. Further away from the closed forest and scattered in the open lake area were trees of *Macrolobium acaciifolium*, the shrubby *Myrciaria dubia,* and a species of *Bactris*.

When the population grows from size category 1 to size category 2 there is a significant mortality which is probably density dependent or caused by abiotic factors since no evidence of pests was seen (Table 3). Plants in size category 1 may be 3–5 years old and they have already been exposed to the dangers of a changing environment (Schlüter 1989). Growth during inundation is close to zero, and growth in the dry periods may be hampered by the drying out of the soil because the seedlings have a poorly developed root system, and because the vegetation, where *A. jauari* grows is relatively open and exposed to the sun. The drying out of the soil, rather than flooding, could therefore be the main factor that determines the mortality of seedlings. High H_2S levels during flooding can, as mentioned above, cause high mortality, but because Laguna Cuyabeno is an *igapó* area such high H_2S-levels are not to be expected here. In the lower Rio Negro area competition for nutrients, variable tolerance to flooding, and removal of seedlings by the water current may in part explain seedling mortality (Piedade 1985).

Mortality in size categories 2, 3 and 4 is low, apparently because these individuals are sufficiently well established to survive the strains of the environment and their growth is faster because the period of total inundation is shorter.

In the Cuyabeno area *Astrocaryum jauari* grows in a narrow band that towards the lake is limited by the failure of seedlings to establish in direct sun and by longer inundations periods and towards the uplands by the dense tree vegetation that produces too much shade for establishment or at least for continued growth.

The vegetation succession along Rio Purus observed by Huber (1906) and described above was apparently determined by available light and degree and length of flooding. Rio Purus is a nutrient rich, whitewater river where erosion and sedimentation are important elements in the formation of the landscape. Laguna Cuyabeno, in contrast, is a blackwater lake, with little erosion and sedimentation, and consequently the morphology of the landscape changes very slowly. Nevertheless there are certain similarities between the two sites, for instance the scattered individuals of *A. jauari* at Cuyabeno which are

Table 3. Numbers of individuals in four size categories of *Astrocaryum jauari* in a 0.1 hectare study plot at Laguna Cuyabeno in Amazonian Ecuador. Size categories: 1. Seedlings and plants with up to four bifid leaves. Most were still attached to the seed and had only two leaves. Plants from this group were mainly found in very dense groups below adult palms. 2. Plants with five or more leaves, all bifid and measuring 45–160 cm in length. 3. Plants with pinnate leaves, but without trunk. 4. Palms with pinnate leaves and trunk.

Size Category	Individuals	% of Size Category 1
1. (Seedlings w. ≤ 4 bifid lvs.)	2640	100.00
2. (Plants w. ≥ 5 bifid lvs., 45–160 cm.)	40	1.51
3. (Plants w. pinnate lvs., without trunk)	42	1.59
4. (Plants w. pinnate lvs., with trunk)	27	1.02

left over from former dense stands, and the present stands that were probably established in the shade of other species. Probably the *A. jauari* stands at Cuyabeno will be stable for a long time, until accumulation of dead organic material has raised the area sufficiently for other less flood tolerant species to take over.

Phenology
Flowering in *Astrocaryum jauari* starts just after the height of the flood season and continues for several months in the Rio Negro area in Brazil (Piedade 1985). The fruits develop during nine months and mature up to the height of the flood season, and the very short fruit dropping period coincides with maximum flooding (or just after) according to observations both in the Rio Machado and Rio Negro areas in Brazil (Goulding 1980, Piedade 1985). At Cuyabeno in Amazonian Ecuador we have a few observations which show that most palms have green fruits and only a few carry mature fruits in the beginning of the flood season.

Pollination
Nothing is published about pollination in *A. jauari*, but inflorescence temperature elevation has been observed in other species of *Astrocaryum* (Henderson 1986, Kraus 1896, Barbosa Rodrigues 1903b) and *Astrocaryum alatum* in Costa Rica has protogyneous flowers, anthesis at night, and beetle pollination (Bullock 1981).

Dispersal
Most fruits of *A. jauari* fall into the water and sink immediately, which explains the dense grouping of seedlings below adult palms, and suggests that water plays a restricted

role in dispersal although there are records of seeds being carried for long distances in running water (Goulding 1980). Many fish species enter the flooded forests to feed on fallen fruits and they are important dispersers of seeds of *A. jauari* and other flood forest species (Goulding 1980, 1983). In Brazil *Colossoma macropomum* and *C. brachypomum* wait underneath *A. jauari* trees where they capture most fruits before they sink. They are able to crush the hard endocarp, and therefore act as predators, but occasionally whole seeds are swallowed, and these are able to germinate when removed from the lower intestine. Other fish, for instance a species of *Brycon,* are known to act as predators as well as dispersers of fruits of *A. jauari,* whereas *Phractocephalus hemiliopterus* and *Megaladoras irwinii* are only known to disperse the seeds (Goulding 1980, 1983). Little is known about the occurrence of these fish in Ecuador although many species of *Brycon*, but none of the others mentioned, are included in Ovchynnyk's (1967) list of Ecuadorean fish species and three species of *Colossoma* and *Phractocephalus hemiliopterus* are listed by Vickers (1976) as being caught and consumed by Siona-Secoya indians in Amazonian Ecuador. The Siona indians at Cuyabeno told us that the fruits of *A. jauari* are also consumed by a number of mammals; White-lipped and Collar Peccary (*Tayassu pecari* and *T. tajacu*) act as predators, Grey Barrigudo (*Lagothrix lagothrica*) and several rodents (*Agouti paca, Dasyprocta* sp., *Proechimus* sp., *Sciurus* sp.) consume only the mesocarp, and may act as short distance dispersers which may also be true of aquatic turtles *(Podocnemis).*

Predation
Fish are believed to be the main predators of *Astrocaryum jauari*. In Brazil large stands of *A. jauari* occur in areas where species of *Colossoma* have been over-fished for a long period (Goulding 1983) which suggests that large populations of these fish may be able to reduce densities of the palm. Fruits that fall on the ground, rather than in the water, are attacked by beetles (Piedade 1985) and peccaries may also act as predators.

Actual and potential uses
Fruits. — Seeds of *Astrocaryum jauari* (Fig. 6) are occasionally consumed as a snack by the children of Siona indians in Amazonian Ecuador and, in Peru along the Río Ucayali, mestizos consume the immature endosperm (Mejia C. 1988). The hard endocarp is used for necklaces (Balslev and Barfod 1987). More important and widespread is the use of the fruits as fish bait, a use which has been observed in Ecuador, Brazil, and Peru (Balslev and Barfod 1987, Barbosa Rodrigues 1903a, Bodley and Benson 1979). The widespread use of fruits as bait, and the findings by Goulding (1980) that many species of fish eat *A. jauari* fruits, suggests that they could be used for fodder in fish farming.

Fresh endosperm contains 15–26% oil (Piedade 1985, Pesce 1985) and dry endosperm 35.2% oil (analyses by V. K. S. Shukla, Aarhus Olie, in 1988). The fatty acid composition of the endosperm in percent of total fatty acid content is as follows: lauric 44.2%, myristic 27.8%, palmitic 7.3%, stearic 3.0%, oleic 10.7%, and linoleic 4.3%.

Figure 6. The fruits of *Astrocaryum jauari* are ovate, up to four centimeters long, glabrous, and yellow when mature. They are an important food for several fish species and are used by humans for a variety of purposes including food, making trinkets, and as a fish bait.

Oil content is 3.2% in the peri/mesocarp and the protein content is 3.2% in the fresh endosperm as well as in the peri/mesocarp (Piedade 1985). Formerly a small oil industry, based on *A. jauari* fruits, was located at Barcelos on the Rio Negro (Lleras and Coradin 1988) and Barcelos is now a center for exploitation of *A. jauari* palm heart.

Trunks. — All species of *Astrocaryum* have hard and durable wood which is used for house construction and, in the past, for making walking sticks and parasol handles (Barbosa Rodrigues 1903a). The wood from *A. jauari* resists rot and is used for construction in Peru (Bodley and Benson 1979, Mejia C. 1988).

Leaves. — A strong white fiber can be extracted from the leaves of *Astrocaryum jauari,* though exploitation has been hindered by its spines and forbiddingly unfavorable habitat according to Schultes (1977) but the Siona indians of Amazonian Ecuador told us that the fibers are not strong. In Peru hats and fans are woven from the young leaves and baskets and sieves from the outer layer of the leaf rachis (Mejia C. 1988). The palm heart is edible and it is also used to treat the pricking pain caused by sting rays (Barbosa Rodrigues 1903a). Commercial exploitation of the palm heart has been initiated along the Rio Negro in Brazil where a processing plant is located in Município de Barcelos (Piedade 1985, Schlüter pers. comm.). Further details about palm heart processing are given in Chapter 11 under *Euterpe chaunostachys*.

Management

At present the only known commercial exploitation of *A. jauari* is the mentioned harvest of palm heart from natural stands along the Rio Negro. Palm hearts are harvested by felling the palm and such large scale exploitation reduces the number of individuals. The resulting reduction in fruit production may influence the fishery in the area significantly. As the palm, especially during its establishment phase, grows very slowly (Schlüter pers. comm.) clearing of large areas will have a long term effect.

Detailed knowledge of the role of *A. jauari* in the vegetation successions in different areas could be used in planning a less destructive harvest of palm heart. *Astrocaryum jauari* is often eliminated by natural processes at a certain phase of the succession on old river bends (Huber 1906) and cutting the palms at that point would produce relatively little disturbance. This might work in theory, but the most easily accessible palms are the ones most likely to be cut.

The palm needs high light intensities to produce fruits (Piedade 1985) so if managed for fruit harvest, selective thinning of other vegetation could favor production and promote regeneration by providing lighter a more open, but not fully exposed, understory.

Cultivation. — Given its limited importance it is unlikely that *A. jauari* has ever been cultivated. There are no agronomic reasons why *A. jauari* should not prosper better in a drier habitat (Schultes 1977) and since seedling growth is slow during inundations it may even grow more rapidly if cultivated on humid *terra firme*. The fact that it grows under widely different degrees and periods of flooding suggests that flooding is not necessary for growth but more likely flooding is a factor that can be tolerated. Floods probably play an important role in triggering flowering and fruiting and it is difficult to predict what would happen to the phenology of *A. jauari* if it was grown on non-flooded land. It could result in year round fruiting, or a reduced production due to pollination failure resulting from not synchronized flowering.

Seedlings for cultivation could be obtained from natural stands. If competition causes seedling mortality (Piedade 1985) thinning of the very dense groups of seedlings might even benefit the regeneration in the stands. The seedlings must be planted in a light open area but not in direct sun. Later, when they are well established, they need more exposure to light. As with many other species of palms, early exposure may induce early flowering. Young plants can be planted in the shade of older palms, where they will be ready to replace these if they are cut for timber or palm heart.

Pests. — Fruits are eaten by a number of animals but otherwise no signs of pests in seedlings and adult palms have been recorded.

Production. — A single tree of *A. jauari* produces 20–40 kilograms of fruits annually of which 40% is endosperm, and of this 14.7–25.5% is oil (Goulding 1980, Piedade 1985, Pesce 1985). From this we estimate that one palm produces 1.2–4 kilograms oil per year which coincides with Lleras' and Coradin's (1988) estimate of an annual production of 2.1 kilograms per palm. At Cuyabeno 27 trunked individuals were found in a 0.1 hectare study plot (Table 3), suggesting that a one hectare natural stand may produce from 318 to 1080 kilograms seed oil per year. The oil resembles seed (kernel-) oil from *Elaeis guineensis*, in that its main fatty acids are lauric, myristic, and oleic. Kernel oil production from *E. guineensis* differs significantly among its varieties but an estimated average is 400 kilograms per hectare and year (Ferwerda 1984). The main purpose of growing *E. guineensis* is, however, to obtain mesocarp oil and not kernel oil.

We have not found published records of yield of palm heart of *A. jauari*.

Harvest. — The trunk of *A. jauari* is heavily armed with spines and climbing is therefore difficult, though methods developed for *Bactris gasipaes* could be used. The peduncle is fibrous but not very hard to cut, and fruits can be harvested with a pole mounted with a curved knife. In inundated forests a boat or raft will often be needed, but, since flooding reduces the distance to the crown, harvesting may prove easy. If grown on *terra firme* fruits could be collected from the ground. The thin, rather dry, mesocarp does not deteriorate quickly, and if the fruits are collected and dried they could be stored.

To harvest palm hearts the palm must be felled, and the palm heart must be consumed or canned within short time because it deteriorates quickly.

Potential in different land-use systems

Cultivation has not been tried, and experiments are needed to show the real potential of *A. jauari* as a crop plant. As a cash crop *A. jauari* will face strong competition; grown for palm heart it will compete with natural stands of *Euterpe* and *Bactris gasipaes* plantations and the slow growth of it will prohibit this use, grown as an kernel oil crop it will compete with *Elaeis guineensis*. It is possible that the equipment used to shell and extract kernel oil from *E. guineensis* can be used for *A. jauari*, which would reduce the investments needed and make it more attractive. In general, however, palms that provide kernel oil are more suitable for harvest from wild stands than are palms that provide mesocarp oil. The seeds are protected by the endocarp which allows the farmer or extractor to store them for long periods and he may therefore, over a period, collect sufficient fruits to justify the trouble and cost involved in bringing them to a merchant.

Extractivism. — In Ecuador stands of *A. jauari* are located in remote areas, they are rather small and scattered, and commercial cropping for palm heart will therefore not be feasible. Given the slow regeneration, and the negative influence cropping for palm hearts would have on the fishery in the area, exploitation would be short sighted. Fruit harvest may also influence the fishery, but, if it is done without felling, the impact is more easily reversible. The short fruiting season of *A. jauari* makes harvest in remote areas easy, because, if it had a long fruiting season, only small amounts of fruits could be obtained at any given time. On the other hand a limited fruiting season means that harvest cannot be postponed if it interferes with other activities. *Astrocaryum jauari* appears to be most common in *igapó*, which, in contrast to *várzea*, is not suitable for agriculture and extractive harvest of it will therefore not conflict with agriculture.

Agroforestry. — In agroforestry systems *A. jauari* would have the advantage of being a tree crop, apparently free of pests, and able to grow on a variety of soils. A disadvantage would be that it can not produce fruits if it was to be grown in the shade of other trees. Small scale cultivation of *A. jauari* for commercial palm heart production would probably not be feasible because it would demand nearby canning facilities. Growing *A. jauari* as a source of fish fodder could be valuable. Fish farming as an integrated part of small scale farming systems can supply farmers with inexpensive and good quality protein but fish farming is not common in the Amazon because of the high number of large rivers which have, so far, produced enough fish to satisfy the demand. The growing demand for fish has, however, resulted in excessive exploitation of this resource in many areas (Goulding 1981) which, in some areas, has made it worthwhile to make fish ponds. Several of the fish species that feed on *A. jauari* fruits are in demand for consumption: *Colossoma macropomum* was the most important fish on the market in Manaus in 1975, and *C. brachypomum* was number five (Goulding 1981). According to U. Saint Paul (pers. comm.) *C. macropomum*, which is already used in fish ponds in Brazil, is well suited for this because it can live in water with low oxygen concentrations such as those often encountered in stagnant water. Young *Colossoma macropomum* consume a variety of plant products, and, until they reach a certain age, they are not able to crack the endocarp of *A. jauari* fruits. Adult *Colossoma macropomum* mainly consume fruits.

Plantations. — No signs of pests have been observed or reported from natural stands of *A. jauari*, and it is likely that it can be grown in plantations. However, competition from other more productive species may be prohibitive to large investments.

Conservation status

Astrocaryum jauari is widespread in northern South America. Even if harvest for palm heart can become a serious threat to the palm in many areas, it is not endangered at the species level. In some areas it may benefit from the impact of man as suggested by Goulding (1983) who found large stands of *A. jauari* in areas where seed predators (*Colossoma* spp.) had been over-fished for a long time. The many fish species that act as dispersers will help *A. jauari* to re-colonize over-exploited areas. The present palm heart exploitation may threaten varieties of *A. jauari* but nothing is known about intraspecific variation, ecotypes, *etc.* Because of its efficient means of dispersal the populations of *A. jauari* in the Amazon basin may be homogeneous although the isolated populations in the Guianas could prove distinct from the Amazonian populations. The International Union for Conservation of Nature (IUCN) lists *Astrocaryum jauari* as not threatened (Dransfield *et al.* 1988).

Research

Francis Kahn from ORSTOM has begun a taxonomic revision of *Astrocaryum* (Coradin and Lleras 1988). Henrik Balslev (Botanical Institute, Aarhus University, Denmark) is carrying out systematic work on *Astrocaryum* for the Flora of Ecuador. Ursula-Birgitte Schlüter from Botanisches Institut on Christian-Albrechts Universität in Kiel, Germany, is working on physiological adaptations to flooding in *A. jauari* (Schlüter 1989). In Brazil Maria Teresa Fernandez Piedade from the Instituto Nacional de Pesquisas da Amazônia (INPA), Fundacão Universidade do Amazonas (FUA), studied the ecology and biology of *A. jauari* (Piedade 1985). At EMBRAPA/CENARGEN (1986) the following are listed as researchers on economic botany, management and domestication of *Astrocaryum:* D. Arcoll (CTAA / EMBRAPA), Peter John Griffee (Grupo Ultra), and Rubens de Oliveira, Jr. (Grupo Ultra).

Specimens studied

The following herbarium specimens were seen or collected for this study. For each specimen the following information is given: *Collector and number (in italics)*, locality, vegetation, elevation above sea level, date, and the acronyms of the herbaria where the specimens are deposited.

Balslev, H. no. 4817. Ecuador, prov. Napo. Reserva Faunistica Cuyabeno, at the northern side of Laguna Grande (76° 11'W, 00° 00'N). Tropical rainforest near blackwater swamps and lakes. 20–26 Jan 1984 (QCA, QCNE, Latinreco, K, NY, AAU); *Balslev, H., Coello, F. and Asanza, E. no. 4313*. Ecuador, prov. Napo. Reserva Faunística Cuyabeno. Laguna Cocodrilococha (76°13'W; 00°02'S). Swamp forest temporarily inundated with blackwater in tropical rainforest. 230 m. 22 May 1983 (QCA); *Henderson, A. and Nascimento de Lima, J. R. do. no. 605*. Brazil. Roraima, Ilha de Maroiá, Mun. Alto Alegra (61°26'W; 3°24'N). Large colonies along river. 15 Jul 1986 (NY); *Pedersen, H. B. no. 67317*. Ecuador, prov. Napo. Laguna Cuyabeno, Laguna Grande, near PUCE field station (76°13'W; 00°00'N). Area regularly inundated by blackwater, 16 Apr 1988 (AAU, QCA, QCNE).

8. MAURITIA FLEXUOSA

Mauritia flexuosa is distributed throughout northern South America east of the Andes where it often forms large stands on acidic, waterlogged soils. It provides many products, including edible fruits, fibers from the leaves, and starch and sap from the trunk. The economic importance varies between geographic regions: in Ecuador no products of it are regularly marketed, whereas in Iquitos, Peru, an elaborate economic network based on its products has developed. *Mauritia flexuosa* is suitable for extractivism and for cultivation and reforestation in swamp lands, but it is largely neglected and deserves more attention from researchers.

Taxonomy

The genus *Mauritia* belongs in the large and diverse palm subfamily Calamoideae, in which all species have fruits that are covered with scales. Balick (1981a) combined *Mauritiella* and *Mauritia*, but Uhl and Dransfield (1987) maintain the two genera as separate based on the following characters: species of *Mauritiella* are multi-stemmed, have root spines on the trunks, have pistillate flowers borne on short branches, and solitary staminate flowers whereas species of *Mauritia* are solitary, unarmed, have pistillate flowers borne singly or on very short branches, and the staminate flowers are in pairs.

Mauritia needs to be taxonomically revised, but probably it includes only two closely related species: *M. flexuosa* [syn. *M. vinifera, M. minor*] with entire leaf sheath margins, and *M. carana* with leaf sheath margins that split into a fibrous cover (Henderson and Balick 1987, Wessels Boer 1965). Only *M. flexuosa*, originally described by Linné f. in 1782 based on a collection of Dalberg from coastal Surinam (Wessels Boer 1965), has been recorded from Ecuador.

Vernacular names

In Ecuador *Mauritia flexuosa* is known as *Acho, Aguaschi, Canangucho, Kanango"cho, Morete,* and *Ne'e'*. The following list gives the references for these names as well as vernacular names from other countries. In parentheses are given the area where the name is used or the language in which it is used, and the literature reference or herbarium specimen from where the information was obtained. Names in bold are the ones most commonly used.

Acho (Ecuador, Balslev and Barfod 1987); *Achuál* (Peru, Nogueira and Machado 1950); *Achul* (Peru, Perez-Arbelaez 1978); *Aeta* (Guyana, Nogueira and Machado 1950); *Aguage* (Peru, Nogueira and Machado 1950); *Aguaje* (Peru, FAO, CATIE 1984); *Aguaschi* (Ecuador, Perez-Arbelaez 1978); *Aguashi* (Ecuador, Sánchez-Monge y Parellada 1980); *Arbol de la vida* (Warrau, Guiana, Johnson 1986); *Awuara* (French Guiana, Pinheiro and Balick 1987); *Bâche* (French Guiana, Nogueira and Machado 1950); *Biriti* (Brazil, Nogueira and Machado 1950); *Boriti* (Brazil, Pinheiro and Balick 1987); *Bororo* (Cuciguias, Bolivia, Dahlgren 1936); *Bruti* (Brazil, Nogueira and Machado 1950); *Buri* (Brazil, Dahlgren 1936); **Buriti** (Brazil, FAO, CATIE 1984);

Buriti do Brejo (Brazil, Pinheiro and Balick 1987); Buritisol (Brazil, MacBride 1960); Burity, Burity do Brejo (Brazil, Nogueira and Machado 1950); Bury (Bahia, Brazil, Nogueira and Machado 1950); Cananguchi (Colombia, Amazonas, Perez-Arbelaez 1978); Canangucho (Spanish, Ecuador, Lescure et al. 1987); Carandéhy-Guassú (Brazil, Nogueira and Machado 1950); Carandá-guassú Carandahi-guassú (Brazil, Pinheiro and Balick 1987); Carandai-guaçu (Brazil, Raulino R. 1974); Cátsú (Mainas, Peru, Dahlgren 1936); Chivaoca (Painonecas, Bolivia, Dahlgren 1936); Chonuya (?, Perez-Arbelaez 1978); Co´nena (Uitoto, Peru, Dahlgren 1936); Coqueiro Buriti (Brazil, Pinheiro and Balick 1987); Eta (Guyana, Nogueira and Machado 1950); Gae-be (Venezuela, Braun 1968); GWY (Macusi, Guiana, Dahlgren 1936); Hiteui (Yakuna-Arawak, Dahlgren 1936); Huaich (Chiquitos, Bolivia, Dahlgren 1936); Ibi-ovi (Peru, Dahlgren 1936); Ideui (Colombia, Dugand 1940); Inahabóto (Guahibo, Venezuela, Dahlgren 1936); Iñéjhe (Bora, Peru, Treacy and Alcorn no. 483); Ino (Muinane-Ges, Peru, Dahlgren 1936); Ita (Guyana, Nogueira and Machado 1950); Ité (Surinam, Wessels Boer 1965); Itéuina (Arútana-Arawak, Dahlgren 1936); Iyō (Puináve, Dahlgren 1936); Izéui (Baré-Arawak, Dahlgren 1936); Kamoí-elg (Arekuna-Carib, Venezuela, Dahlgren 1936); Kanango"cho (Kofan, Ecuador, Lescure et al. 1987); Koaie (Auaké, Venezuela, Dahlgren 1936); Koj (Surinam, Wessels Boer 1965); Kuése (Schirianá, Venezuela, Dahlgren 1936); Kuia (Yekuana, Venezuela, Braun and Delascio C. 1987); Kui (Wayumará, Dahlgren 1936); Liōkoho (Xiriana-teri, Venezuela, Anderson 1978); Marití (Colombia, Dugand 1940); Maurisie (Surinam, Wessels Boer 1965); Mburity, Merity, Mirití (Brazil, Nogueira and Machado 1950); Miritizeiro (Brazil, Pinheiro and Balick 1987); Mirity (Brazil, Nogueira and Machado 1950); Mirutí (Brazil, Wallace 1853); Moréte (Quichua, Ecuador, Balslev and Barfod 1987); Morisi (Surinam, Wessels Boer 1965); Moriti (Brazil, Pinheiro and Balick 1987); Mority (Brazil, Nogueira and Machado 1950); Moriche (Colombia, Venezuela, Braun 1968); Muriche (Venezuela, Braun 1968); Murichi (Brazil, Nogueira and Machado 1950); Muriti (Brazil, Pinheiro and Balick 1987); Murity, Murity do Brejo (Brazil, Nogueira and Machado 1950); Muritiseiro (Brazil, Pinheiro and Balick 1987); Murityseiro (Brazil, Nogueira and Machado 1950); Nanicuni (Colombia, Coreguaje, Perez-Arbelaez 1978); Ndesé (Yagua, Peru, Dahlgren 1936); Ne' e´ (Siona, Ecuador, Lescure et al. 1987); Neéyo (Tucano, Dahlgren 1936); Nesicaná (Peru, Dahlgren 1936); Nxoroti (Iquito-Cahuaranu, Peru, Dahlgren 1936); Ocon (Quitemocas, Bolivia, Dahlgren 1936); Oheed (Warrau, Guiana, Dahlgren 1936); Ohido (Warao, Venezuela, Johnson 1986); Ovi (Peru, Dahlgren 1936); Ovino (Brazil, Dahlgren 1936); Palmier Bâche (French Guiana, Nogueira and Machado 1950); Paníhibe (Baré-Arawak, Dahlgren 1936); Papao (Chapacúras, Bolivia, Dahlgren 1936); Shispi (Chayahuitas, Peru, Dahlgren 1936); Sosiqui (Cayavavas, Bolivia, Dahlgren 1936); Téui (Baniwa-Arawak, Dahlgren 1936); Teuíra (Tariana-Arawak, Dahlgren 1936); Tióg (Makú, Brazil, Dahlgren 1936); Tomeao (Muchogéonés, Bolivia, Dahlgren 1936); Tsil(e)ká (Makú, Dahlgren 1936); Tsípara (Chebero, Peru, Dahlgren 1936); Quiteve (Venezuela-Brazil, Sánchez-Monge y Parellada 1980); Uara (Venezuela, Sánchez-Monge y Parellada 1980); Uque (Paunacas, Bolivia, Dahlgren 1936); Wá-la (Yahuarána-Carib, Venezuela, Dahlgren 1936); Viño (Pano, Nakoman, Peru, Dahlgren 1936); Vinon (Shipibo, Peru, Bodley and Benson 1979); Yelasu (Nambicuara, Brazil, Johnson 1986).

Morphology

Mauritia flexuosa is a solitary, large sized palm with a trunk up to 40 meters tall and a dbh of 30–60 centimeters (Fig. 7). Its trunks are clean and when young they have conspicuous nodes. The costa-palmate leaves are erect and later spreading, numbering

Figure 7. A stand of *Mauritia flexuosa*, growing at the margin of a blackwater oxbow lake in Amazonian Ecuador. The palm is up to 40 meters tall with trunks up to 60 centimeters in diameter, and the costapalmate leaves up to six meters long, and their blades up to two and a half meters across.

8–12, the blade has a radius of about 2.5 meters, the entire leaf is more than six meters long, and often 5–8 dead leaves hang from the crown. *Mauritia flexuosa* is dioecious with up to eight interfoliar inflorescences, each measuring more than two meters and with 25–40 pendant rachillae from the horizontal or pendulous main axis. The fruits are globular to ovate, 3–7 centimeters long, with reddish scales that cover the yellow mesocarp and one large seed with white homogeneous endosperm. The root system of a *M. flexuosa* growing in acidic humus in a savanna was studied by Granville (1974). The main root-mat grew 20 centimeters below the ground, reaching a horizontal distance of 40 meters from the palm, and covering an area of more than 5000 square meters. From this horizontal layer secondary roots grew upwards, the part below the soil surface having third order roots which again branched into fourth and fifth order roots. The secondary roots carried pneumatophores, active in gas exchange, above the soil surface. Our colleague, Finn Borchsenius, has observed mycorrhiza on *M. flexuosa* in Ecuador.

Distribution

Mauritia flexuosa is distributed throughout northern South America east of the Andes. In Ecuador it is common in the Amazonian lowland up to about 800 meters above sea level (Fig. 8). We have observed it in cultivation up to 970 meters near Zamora (see climatic diagram 2 in Chapter 12). In Ecuador it occurs only in areas where precipitation exceeds evapotranspiration in every average month, but it may occur in areas with a pronounced dry season if water is available throughout the year, such as in some areas of the Llanos in Venezuela and near Belém in Brazil (climatic diagram E and D in Chapter 12). Scattered individuals indicate local swampy land or a yearly rainfall of two meters or more (Moore 1973b). Mean annual precipitation in its distribution area varies from 1141–6315 mm, mean annual temperatures varies from 22.8–27.1°C. Climatic diagrams A, D, E, 1, 4, and 5 in Chapter 12 are from localities where *M. flexuosa* occurs.

Habitat

In the large savannas of northern Bolivia, eastern Colombia, Venezuela, and the Guianas *M. flexuosa* grows in gallery forests along rivers and scattered or in stands in wet depressions, often with a herbaceous undergrowth of various species of Cyperaceae and Poaceae, and on soils that are acidic and rich in organic material (Aristeguita 1968, Blydenstein 1967, Granville 1974, Moraes R. 1989, Sarmiento 1983, Wessels Boer 1965). In swampy areas in tropical rainforest *M. flexuosa* is often found in extensive mono-specific stands or together with other palms such as *Jessenia bataua* and *Euterpe precatoria* (Kahn 1988). A survey based on remote sensing showed that 21% of a 310,000 hectare area near Iquitos was occupied by *M. flexuosa* and millions of hectares of such *Aguajales* occur in eastern Peru (ONERN 1977, Salazar C. 1967). Surveys along Río Itaya near Iquitos by Salazar and Roessl (1977) showed an average of 246 trunked individuals per hectare (based on 10 x 0.5 hectare), and Gonzáles (1971–1974 cited in Kahn 1988) found an average of 351 trunked individuals and 297 juveniles without trunks per hectare in the upper Huallago valley in Peru (based on 20 x

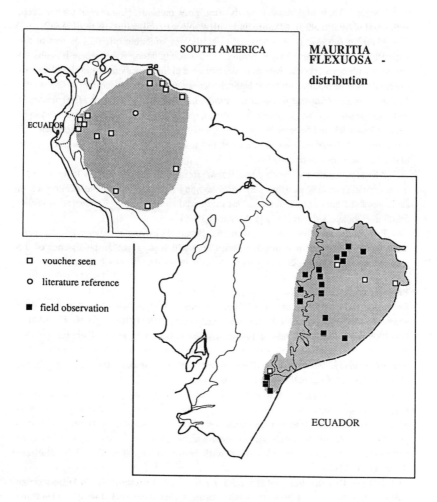

Figure 8. Distribution of *Mauritia flexuosa* in South America and in Ecuador. Its total distribution covers the Amazon basin. In Ecuador it is restricted to the lowlands east of the Andes. The sign for 'vouchers seen' shows the collecting localities for the herbarium specimens cited at the end of this chapter. The sign for 'literature reference' show sites which were given in the literature with such exactitude that they could be placed on the map. The sign for 'field observations' shows such localities where we have observed, but not collected the palm.

0.05 hectare). These high density stands often grow on acidic peat several meters deep, composed of litter from *M. flexuosa* and with a soil-water pH of 3.5 (Kahn 1988).

Mauritia flexuosa is not restricted to acidic peat soils but often forms part of the vegetation succession on newly deposited sand and mud along rivers where it is able to compete with other species. Stands of *M. flexuosa* along small streams in the Llanos of Colombia change rapidly into broad leaf semi-evergreen forest (Blydenstein 1967). In the Amazon estuary *Mauritia flexuosa* is found in a zone behind belts of *Montrichardia* and *Euterpe oleracea* and together they form a mechanism for colonizing mud and sand banks (Strudwick and Sobel 1988, Richards 1979). Upriver from the Amazon delta *M. flexuosa* is not found on the river margins but in eastern Peru, *M. flexuosa* is common in old oxbow lakes (Moore 1973b).

Alone, or together with other species, *M. flexuosa* forms the canopy layer where it occurs. It regenerates well in open areas judging from its widespread occurrence in savannas and from our own observations in cleared swamp forest. Regeneration is also found in complete and partial shade (Richards 1979).

In Ecuador we have not seen *Mauritia flexuosa* growing spontaneously on *terra firme* but in cultivation it seems to thrive there, which suggest that the absence of the palm from well drained soils is due to competition rather than to abiotic factors.

Phenology

Flowering and fruiting appear to be unevenly distributed, but both occur throughout the year. In the Colombian Amazon most individuals of *M. flexuosa* flower almost synchronously, starting in July and August (Urrego G. 1987) which coincides with the end of the main rainy season, maximum flooding, minimum sunshine hours per day, minimum average temperature, and minimum difference in daily temperatures. In the Orinoco delta in Venezuela there are two flowering seasons for *M. flexuosa*, one with mostly staminate flowering in May which is triggered by rain showers that proceed the main rainy season, and one with mostly pistillate flowering in December which is triggered by a smaller increase in precipitation (Heinen and Ruddle 1974). In the Orinoco population of *Mauritia* starch content in the trunk reaches a maximum at the start of inflorescence formation which agrees with results obtained in Brazil by Barbosa Rodrigues (1903b).

In Colombia the fruits of *Mauritia flexuosa* are immature but full sized seven months after flowering (Urrego G. 1987). In the Orinoco development of mature fruits needs four to five months from the onset of flowering (Heinen and Ruddle 1971). Mature fruits are found all year around, with a marked peak in August through October and a smaller peak in February through April.

Periodic peaks in flowering and fruiting have also been reported from Peru, Brazil, and Ecuador (Mejia C. 1988, Cavalcante 1974, Vickers 1976). During our fieldwork in Amazonian Ecuador we have only rarely observed flowering individuals, but often fruiting ones and according to local people in the Coca area the main fruiting season is January through March.

Pollination

We are not aware of pollination studies of *M. flexuosa* but since it is dioecious, has a conspicuous inflorescences, and produces an intense scent it is probably pollinated by insects (Urrego G. 1987).

Dispersal

Mauritia flexuosa must possess effective means of dispersal, judging from its wide distribution and the occurrence of widely scattered individuals. Humans and animals both contribute to its dispersal; humans carry fruits with them for consumption, and discarded seeds may give raise to palms around settlements (Mejia C. 1988); cultivation of the palm influence its distribution; animals take its fruits but mostly consume only the nutritious and soft mesocarp and leave the hard seed whereby they act as short distance dispersers.

Based on local informants in the Colombian Amazon Urrego G. (1987) listed *Myoprocta acouchy*, *Agouti paca*, *Tayassu tajacu* (Collar Peccary), *Dasyprocta fuliginosa* (Agoutis), and *Tapirus terrestris* (South American Tapir) as consumers of the mesocarp and short distance dispersers, and *Tayassu pecari* (White-lipped Peccary) as a consumer of whole fruits, including the seeds. Monkeys and squirrels also consume the fruits (Perez-Arbelaez 1978). In Ecuador people hunt near fruiting *M. flexuosa* because Peccaries and Agoutis seek their fruits. *Mauritia* fruits are consumed by specialized frugivorous birds and have characters that are typical of fruits dispersed by such birds (Snow 1981); the seed is large compared to the fruit size and it is surrounded by a nutritious mesocarp.

Mauritia flexuosa often grows in inundated areas but there are no records that fish disperse its fruit, although this cannot be excluded. Other fruits, as large as those of *M. flexuosa*, are swallowed whole and probably dispersed by some fish for example the fruits of *Neolabatia* are swallowed by *Colossoma macropomum* (Goulding 1980). The Siona indians of Amazonian Ecuador use the fruit mesocarp for fish bait (Vickers 1976), so apparently it is palatable to some fish species. Water may be responsible for long distance dispersal of *Mauritia flexuosa* according to Ridley (1930) and whole fruits are often found on the beaches in French Guiana, deposited by the sea (Granville 1974). Fresh fruits sink in water, but their seeds float for a few days, if they have been left to dry for a period, but it is not known whether such seeds are capable of germinating.

Actual and potential uses

Inflorescences. — The young unopened inflorescence bud can be cut and tapped for its sap which contain about 50% sugar (Raulino R. 1974) and may be drunk directly, fermented to palm wine, or boiled down to sugar. A more dramatic approach involves felling, defoliating, and scorching the palm with fire in order to stimulate the flow of sap from its inflorescence stalks (Corner 1966).

Table 4. Composition of 100 grams of fresh pulp from *Mauritia flexuosa* (Adapted from Bohórquez R. 1976).

Protein	3.0	g
Fat	10.5	g
Fiber	11.4	g
Ash	1.2	g
Ca	113.0	mg
P	19.0	mg
Fe	3.5	mg
Vitamin B1	0.03	mg
Vitamin B2	0.23	mg
Vitamin C	26.00	mg
Vitamin A	12.38	mcg

Fruit-pulp. — The fruit-pulp (mesocarp) of *M. flexuosa* is nutritious and important in the diet of many indigenous groups in the Amazon (Table 4). In parts of Venezuela the pulp is dried and roasted after which it is less perishable and serves as a kind of bread (Braun 1968). In the Orinoco area, also in Venezuela, fruits, ripe enough to fall by themselves, are collected, their exocarp is rubbed off, and worked into a mash which is then wrapped in leaves, "and enveloped in a framework of slips of blowing-cane palm, made first into a cylinder, and then the ends brought together and tied tightly, so as to bring it to a spindle-shape" (Spruce 1908 [1970], cited in Beckermann 1979). The pulp can be kept like this for weeks, and will start to ferment; later it is mixed with water, the scales are removed, and it is ready to drink. Beckermann suggested that fermentation may increase the protein content in the pulp owing to multiplication of the microorganism responsible for fermentation. Among Yanomama indians in Brazil oil is extracted from *Mauritia flexuosa* by boiling the fruits and skimming the oil off (Anderson 1978) and in Iquitos, Peru, preserves and ice cream flavor are made from the fruits (Pesce 1985, Blicher-Mathiesen pers. comm.).

In Ecuador *Mauritia* fruits (Fig. 9) are important to all the indigenous groups in the Amazon lowlands and to the Siona-Secoya indians it is one of the most important wild fruits (Vickers 1976). They fell the palm, bury the immature fruits in the ground for some days, and use the major part of the mature fruits to make a drink called *chicha*. For this the fruits are soaked and mashed in water, using a stick or the hand and the resulting milky fluid is filtered and consumed without further preparation, or sugar, honey, banana mash, or some other source of carbohydrate is added, and the drink is left to ferment for a day or two. For direct consumption the fruits are soaked in lukewarm water, the scales are peeled off, and the mesocarp eaten.

Trade with fruits or fruit products of *M. flexuosa* has not been recorded for

Ecuador, and little has been published on the matter from other areas, Iquitos in Peru being the exception. Padoch (1988) gives a detailed description of the significant role that these fruits play in the economy of that area: *Mauritia* fruits are used in a variety of products, including ice cream, frozen desserts, juice, and for direct consumption. A complicated system involving harvesters, wholesalers, retailers, and processors has emerged, and, although export from the area is limited, a small amount of fruits is sold to Japan to be used in ice cream (Blicher-Mathiesen pers. comm.). Given the growing demand for exotic products in Europe and North America canned or fresh fruits of *Mauritia flexuosa* could represent a potential resource for export.

Seeds. — The seeds, which make up 54% of total fruit weight and consist of manu-cellulose, are sometimes used for buttons and other small things (Pesce 1985, Perez-Arbelaez 1978, Barbosa Rodrigues 1903b) but even though they appear to be suitable for this purpose they are normally discarded after the fruit pulp has been used. The manu-cellulose can be transformed into fermentable sugar through acid hydrolyses or with the use of diastase, an enzyme found in germinating palm seeds (Pesce 1985). Another possibility would be to mill the seeds and use them for cattle fodder in the same way as mentioned for *Phytelephas aequatorialis* (Chapter 3). Figures from Iquitos can illustrate the unused potential of these seeds; the daily consumption of *M. flexuosa* fruits in Iquitos is estimated at 15 tons (Padoch 1988), and using Pesce's data, 54% or 8.1 tons is seeds, thus almost 3000 tons of seeds are discarded every year from Iquitos alone.

Trunks. — Up to 500 edible larvae of the beetle *Rhynchophorus palmarum* may live in a single decaying trunk of *M. flexuosa* and both *colonos* and indigenous people chop trunks open to facilitate access for the beetles and increase their oviposition. The larvae are occasionally sold on the market in Coca in Amazonian Ecuador and it is the only *M. flexuosa* product which, to our knowledge, is marketed in Ecuador.

The trunks are only rarely used for construction because the pith is soft, but occasionally planks are made by splitting off the hard outer layer. In Brazil canoes are made by hollowing out the trunk (Pinheiro and Balick 1987).

Sap for wine has been collected from *Mauritia flexuosa* by many indigenous groups, usually by making holes into a felled trunk or into the stump (Braun 1968). In one day a single palm can yield 8–10 liters sap which contains mainly water and sucrose and can be used for making wine or sugar (Pesce 1985). *Mauritia vinifera* was so named because of this practice.

Starch (sago) is produced in the trunk of *Mauritia flexuosa*. This product is extracted from a number of palm genera of which *Metroxylon* from Southeast Asia is the best known example. In South America *Mauritia flexuosa*, *Syagrus romanzoffianum*, *Manicaria saccifera*, and *Roystonea oleracea* have been used for extraction of starch, but only on a local scale (Ruddle *et al.* 1978).

Figure 9. Infructescence of *Mauritia flexuosa*, collected from a low palm, about six meters tall (seen in the background). The rachis emerges in the leaf axils and the weight of the fruits, attached to the pendant rachillae, bend it down.

Table 5. Composition of pith starch in % of the dry weight from a staminate *Mauritia flexuosa* following hydrolysis. Samples taken 1 meter and 13.2 meters above ground. Voucher specimen: *Pedersen no. 67310.* The pith was fragmented and methanolized, trimethylsilylized, and chromatographed. The analysis does not give information about how the sugars are bound (analyses by Peter Fromholt, Grindsted Products A/S, Denmark.)

Height above ground	1 m % dw	13.2 m % dw
Rhamnose	0.2	<0.1
Arabinose	1.3	2.2
Xylose	5.6	4.4
Galacturonic acid	3.2	2.5
Galactose	0.8	1.4
Glucose	12.0	9.0

Among the Warao indians, who live in the Orinoco delta in Venezuela, starch from *M. flexuosa* is an important source of storable carbohydrates. Ruddle *et al.* (1978) described the way they extract it: first a suitable palm (one which has not yet developed flowers that season) is localized, its starch content is checked by chopping an axe into the trunk and examining the blade, if the test is positive the palm is felled and the soft spongy pith is chopped into pieces, the pith is then placed on a sieve and hand kneaded while watered to remove the starch, below the sieve the water is collected in a small container inside a larger one, the starch settles on the bottom of the inner container, excess water flows over to the outer to be re-filtered several times. The resulting starch may be stored as flour or formed into bricks and baked on stones or iron plates (Lévi-Strauss 1952). In some parts of South America the starch is used medicinally as a remedy against dysentery and diarrhoea (Plotkin and Balick 1984). While species of *Metroxylon* often yield more than 200 kilograms of starch per trunk, the average yield of *M. flexuosa* is only 60 kilograms per trunk (Ruddle *et al.* 1978) but only a few individuals have been examined and higher yielding varieties may exist. According to the literature only pistillate palms are exploited for starch by the indigenous populations. We were unaware of this when sampling a staminate individual for analysis of starch composition, so our results, shown in table 5, should be read with this in mind.

Leaves. — Salt has been obtained by burning the leaves of *Mauritia flexuosa* and boiling the ash until only a brown powder is left (Lévi-Strauss 1952). From the petioles a spongy material can be obtained which is used to cork bottles, or, made into strips, the spongy material is woven into mattresses or sitting mats, a use we observed among Quichua indians in Amazonian Ecuador and Bodley and Benson (1979) reported it from Peru. The petioles may also be used for paper production (De los Heros G. and Zárate

1980–1981). The leaf blades of *M. flexuosa* are occasionally used for thatch, and from the leaf bud string is obtained. Wallace (1853) described the way string is made: "This [the leaf bud] is cut down, and by a little shaking the tender leaflets fall apart. Each one is then skilfully stripped of its outer covering, a thin ribband-like pellicle of a pale yellow color which shrivels up almost into a thread. These are then tied in bundles and dried, and are afterwards twisted by rolling on the breast or thigh into string, or with the fingers into thicker cords." The cord made from young buds is extremely pliable and durable (Dahlgren 1944) and a less durable strings can be made from older leaves (Schultes 1977). The strings are used for hammocks, ropes, and fishing nets. In Brazil, mainly in the states of Maranhão and Pará, there was a production of 614 tons of fiber from the leaves of *M. flexuosa* in 1980 (Balick 1984). The palm heart is edible (Bohórquez R. 1976, Calzada 1980).

Management

In Peru seedlings of *Mauritia flexuosa* from nearby swamps are sometimes used for planting it as an ornamental or for its fruits and other products (Calzada 1980). In Brazil the Kayapó indians plants *M. flexuosa* in their fields, and harvest them during the long fallow period that follows the period of intensive cultivation (Posey 1984). Thus cultivation does occur, but it is limited and fruits and other products are usually obtained from wild palms.

The only management which is commonly applied is felling, and around Iquitos in Peru this has seriously depleted most stands within a days travel from the town (Padoch 1988). Since pistillate individuals are the ones most commonly felled, heavy exploitation can influence the sex ratio and in exploited stands in the lower Río Ucayali valley above Iquitos three stands have a predominance of staminate palms (Kahn 1988) but variation may be large and in Colombia Urrego G. (1987) found pistillate dominance (83%) in some plots and staminate dominance (77%) in others.

Sustained yield management of natural stands involves non-destructive harvest methods and thinning of other vegetation. Favoring pistillate palms by selective thinning of staminate, could be employed, but the optimal sex ratio is not known.

Cultivation. — There is a strong negative correlation between germination rate and seed age (López C. 1968, quoted in Bóhorquez R. 1976): newly harvested seeds have 100% germination after 75 days and seeds sown after 20–30 days have 55% germination after 120 days; seeds which have been stored one week at 5°C and sown after a total of 30 days have 95% germination after 75 days. The palm can be planted on permanently water logged soil as well as on *terra firme,* and it can tolerate very acidic soils but it should not be planted in areas where its pneumatophors will be inundated for long periods because, as discussed above, it can probably not survive this.

If *Mauritia flexuosa* is planted to produce fruits, its dioecious sex system may be a problem. It is, at present, not possible to determine its sex before the first flowering and therefore an excess of male individuals may result. This problem can be reduced by planting in dense stands, which are later thinned. According to Calzada (1980) 3% males is sufficient to ensure pollination, but the basis for this number is not mentioned. It appears low and certainly demands a strict synchronization of flowering. The figures from natural stands suggest that planting densities could be very high, probably at least 400 individuals per hectare.

We have seen mycorrhiza associated with roots of adult *Mauritia*, but nothing is known about the possible beneficial effect of this.

Flowering starts when the palm is about eight years old according to Bohórquez R. (1976) but local people, who grow the palm in Ecuador, told us that flowering starts anywhere after 5–11 years. Trunk formation may start early; for instance we found a 1.5 meters tall trunk in a four-year-old palm.

Pests. — No important pests have been recorded for *M. flexuosa* even though it often grows in large mono-specific stands. Larvae of *Rhynchophorus palmarum*, *Rhinosthomus barbirostris,* and *Metamasius hemipterus* have been observed in the trunks of dead or wounded palms (Urrego G. 1987), but they are not known to cause damage in healthy palms.

Production. — Cavalcante (1974) found 724 fruits in one infructescence on a cultivated palm with eight infructescences suggesting a total of about 5800 fruits on the tree. Using an average weight of 50 grams (Padoch 1988) production can be estimated as 290 kilograms of fruits per tree per season. In Peru a fruit production of 19 tons per hectare was obtained in a plantation with 100 palms per hectare which corresponds to 190 kilograms per palm per season. Lleras and Coradin (1988) estimated a fruit production of 200 kilograms per palm, yielding 24 kilograms of oil. The oil resembles mesocarp oil from *Elaeis guineensis* according to analysis by V. K. S. Shukla, Aarhus Olie, in 1987.

Starch production was estimated above to 60 kilograms per trunk. Production figures for fibers and palm heart do not exist.

Harvest. — Fruits are dark red and start to fall from the infructescence when they are ripe. At this stage they deteriorate rapidly, and if they must be transported they should be harvested earlier. Infructescences may form as low as 2.2 meters above the ground and harvest is then easy using a ladder and a *machete*. This method, or the use of a pole with a curved knife, may be applied for some years, but as the palm grows it becomes increasingly difficult because the infructescences are interfoliar and difficult to reach by climbing. Adequate harvest methods are not known.

Loss of nutrients from the ecosystem through harvest of *Mauritia* fruits can, in part, be estimated by using the figures in table 4. The pulp (mesocarp) and epicarp make up 46% of the fruit (Pesce 1985), and the average N-content in protein is 16% (Schønheyder and Nørby 1965). Loss per hectare with a yield of 19 tons of fruits is about 42 kilograms of N, 9.9 kilograms of Ca, 1.7 kilograms of P, and 0.3 kilograms of Fe. This does not include the kernel, which consist mainly of manocellulose. Since *M.*

flexuosa is mainly found in nutrient poor areas, such a large nutrient export is likely to lead to reduced production unless fertilization or long fallow periods are used.

Harvest of palm heart and sago will kill the palm. Tapping sap may be done in a sustainable way if later infection from *Rhynchophorus palmarum* can be avoided which may be achieved by using a small net to stop beetles from entering during tapping and by blocking the hole after tapping.

Potential in different land-use systems

At present there are many areas were there is little or no demand for the fruits of *Mauritia flexuosa* (Fig. 10) and in order to cultivate *M. flexuosa* as a cash crop, a market must be created. The estimates for its oil production given by Lleras and Coradin (1988) means that, with 200 pistillate *Mauritia* palms per hectare, the production will equal that of *Elaeis guineensis* plantations, so it has a considerable potential as an oil crop. A matter of significant importance would be the commercialization of the seeds for fodder, alcohol production, or vegetable ivory. The seeds can be stored for a long time, which makes it possible for small farmers and extractors to collect sufficient seeds to make the transportation to the merchants worthwhile. Commercialization of dried mesocarp for production of flour would be equally beneficial. Large stands, whether natural, in plantations, or grown by many small farmers can provide sufficient products to make installation of equipment for oil extraction, starch extraction, and milling of seeds worthwhile. Apart from being used in agriculture *M. flexuosa* also offers good possibilities for replanting in cleared swamp areas, given its high regenerative potential in open land.

Figure 10. Fruits of *Mauritia flexuosa,* showing the scales of the epicarp, the soft mesocarp, and the hard endosperm in the center.

Extractivism. — Subsistence extractivism of fruits, sap, starch, and palm hearts of *M. flexuosa* gives significant contributions to the diet of many people in the Amazon and its rapid growth in large swamp areas means that it can easily tolerate this form of exploitation, even when it involves felling the palms. Commercial extractivism of *M. flexuosa* also offers great possibilities considering the large, high density stands of it in many areas. The situation around Iquitos does, however, illustrate the problem of destructive harvesting, which may severely reduce regeneration because of lack of seeds so if harvest is done by felling, the use of starch, fiber, and palm heart should be organized along with it.

Agroforestry. — For sustainable farming in swamp areas, *M. flexuosa* could be a very valuable component, which could supply a number of products. Fruit yield is high and the nutritious fruits can be used for human consumption supplying oil, good quality flour, and fodder for domestic pigs and chickens. Used for fodder the harvest is easy because the animals may eat the fruits when they fall to the ground. Other advantages include, that it can be grown on a variety of soil types, that high palms give very little and diffuse shade, and that no serious pests are known. Among the disadvantages are that *Mauritia* needs a long time to mature and start fruit production and that its extensive and superficial root system may compete with other crops with shallow root systems. However, the root system may also act as a nutrient trap and a soil stabilizer on land prone to erosion.

Plantations. — *Mauritia flexuosa* appears to be free of serious pests, and produces as much oil as *Elaeis guineensis* and it may therefore become a valuable alternative to *E. guineensis* which is seriously attacked by a number of pests. Development of a plantation industry would depend on development of adequate harvest equipment.

Conservation status

Mauritia flexuosa is very common throughout northern South America east of the Andes, especially abundant in swamps which are seldom used for agricultural purposes. New habitats for the palms are constantly created through formation of oxbow lakes and in estuaries, which, in connection with the good regeneration ability observed in cleared areas, leads us to conclude that at present *M. flexuosa* is not threatened, although intensive destructive extractivism may lead to loss of important genetic variation. The International Union for Conservation of Nature (IUCN) lists the status of *Mauritia flexuosa* as not threatened (Dransfield *et al.* 1988).

Research

Mauritia flexuosa has received little attention in research programs. No genetic improvement has been undertaken and almost nothing is known about its ecological requirements, breeding systems, variation in production, and composition of products. In Ecuador a study of *Mauritia* oil has been carried out at the Escuela Politecnica

Nacional in Quito (Pacheco 1989). According to Coradin and Lleras (1988) research on ecology and reproductive biology of *Mauritia* populations has been initiated near Manaus in Brazil by INPA. In EMBRAPA/CENARGEN (1985) the following are listed as researchers on *Mauritia*: Coradin, L. (EMBRAPA/CENARGEN), Kahn, F. (ORSTOM), Lleras P., E. (EMBRAPA/CENARGEN), Mejia, G., M. (U. Nacional, Palmira).

Specimens studied

The following herbarium specimens were seen or collected for this study. For each specimen the following information is given: *Collector and number (in italics)*, locality, vegetation, elevation above sea level, date, and the acronyms of the herbaria where the specimens are deposited.

Alarcón, R. no. 139. Ecuador, prov. Napo. Nuevo Rocafuerte y la Orilla del Río Napo hasta 5 km al Oeste y la orilla del Río Yasuni, hasta la Laguna de Jatuncocha (QCA); *Baker, M. A. no. 6171*. Ecuador, prov. Morona-Santiago. La Mission, 5 km al Sur del Río Bomboiza y cerca la carretera (78°31'W; 3°26'S). 800 m. 14 May 1985 (NY); *Balick et al. no. 921*. Brazil, Amazon. Santarem Cuiaba road km 879, Br 163 from Santarem, Pará. 11 Nov 1977 (NY); *Davidse et al. no. 5683*. Colombia, Caquita. 7 km southeast of Morelia along road to Río Pescado, S.W. of Florencia. (75°41'W; 1°31'N). 10 Jan 1974 (MO); *Davis, W. E. and Marshal, N. no. 1185*. Bolivia, Dep. Beni. prov. Ballivan. Río Chimane. Environs Fatima. 320 m. 8 Jun 1981 (NY); *Gentry et al. no. 28866*. Peru, Loreto. Mishana. Río Nanay, halfway between Iquitos and Santa Maria de Nanay. 140 m. 21 Jul 1980 (MO); *Killeen et al. no. 1409*. Bolivia, Dep. de Santa Cruz. prov. Nuflo de Chavez. Estancia Las Madres 12 km Norte de Concepción. (62°00'W; 16°00'S). 6 Nov 1985 (F, LPB, not seen); *Moore, H. E. Jr. no. 10317*. Surinam. Sandy soil beside W. J. van Blommestein Meer on road to Brownsberg. 3 Jul 1977 (BH); *Moore, H. E. Jr. et al. no. 10355*. Surinam. Way to Cubiti. Wet Savanna on sand. 15 Jul 1977 (BH); *Pedersen, H. B. no. 67310*. Ecuador, prov. Napo. INIAP-Payamino station, near Coca at Coca-Loreto road (77°06'W, 00°28'S). 250 m. Primary swamp-forest; very common in the area. 8 Mar 1988 (AAU, QCA, QCNE); *Steyermark, J. A. no. 87708*. Venezuela, Territorio del Amacuro. Along Río Amacuro above San José de Amacuro. Mangrove and Moriche swamp. 21 Nov 1969 (F). *Steyermark, J. A. no. 57649*. Venezuela, Bolívar. Between Río Caroni and Ciudad Bolivar. 200 m. Savanna. 2 Aug 1944 (NY); *Steyermark, J. A. et al. no. 108392*. Venezuela. Laguna La Bodega immideamente al Este de Santa Fe. (64°24'W; 10°17'N). 17 Sep 1973 (MO, NY); *Steyermark, J. A. et al. no. 114861*. Venezuela. Terr. Fed. Delta Amacuro. Depto. Antonio Diaz. 17 Oct 1977 (MO); *Treacy J. and Alcorn, J. B. no. 483*. Peru, Dist. Pebas. About 180 km east northeast of Iquitos. (72°05'W; 3°00'S). Bora native community of Brillo Nuevo, Yaguasyacu River affluent of Ampiyacu-River. 106 m. 11 Dec 1981 (NY); *Williams, L. no. 12665*. Venezuela, Estado Bolívar. En la sabana de la Sardina, La Paragua. 70 m. 19 Mar 1940 (F).

9. JESSENIA BATAUA

Jessenia bataua is widely distributed in northern South America and has long been appreciated for its fruits which provide food, oil, and the beverage called *chicha*. It has often been described as an oil crop with a large under-exploited potential, and subsistence extractivism and cultivation of *J. bataua* can furnish good quality oil for local consumption. Commercial exploitation, however, is still limited, because of scattered fruit sources and lack of adequate harvest and extraction facilities. Extensive research on *J. bataua* has started in recent years.

Taxonomy

Jessenia belongs to the palm subfamily Arecoideae and is closely related to *Oenocarpus*, the most important feature separating the two being that *Jessenia* has a ruminate endosperm whereas *Oenocarpus* has a homogeneous endosperm. In a recent revision of the *Jessenia-Oenocarpus* complex Balick (1986) synonymized a number of names and *Jessenia* is now considered a monotypic genus with *Jessenia bataua* as the only species. Two subspecies are recognized: 1) *Jessenia bataua* subsp. *bataua* [synonyms: *Oenocarpus bataua*, *Jessenia polycarpa*, *Jessenia repanda*, *Oenocarpus seje*, *Jessenia weberbaueri*] and 2) *Jessenia bataua* subsp. *oligocarpa* [synonyms: *Jessenia oligocarpa*, *Oenocarpus oligocarpa*]. Number and position of pistillate flowers, length of staminate flowers, number of stamens, number and diameter of rachillae, and such leaf characters as gray versus whitish color and many versus few scales on the abaxial side of pinnae distinguish the two subspecies. *Jessenia bataua* subsp. *oligocarpa* is distributed in Trinidad, Guyana, and north and northeastern Venezuela, and subsp. *bataua* is found throughout northern South America and Panama. The original description of *Jessenia bataua* [as *Oenocarpus bataua*] was based on a collection by Martius from Amazonas, Brazil, without collection date and number (Martius 1823).

Vernacular names.

In Ecuador *Jessenia bataua* is known as *Chapil, Cola-boca, Cola-pa-chi, Cosa, Gôsa, Kula'po-tci, Milpesos, Shimpi, Shigua*, and *Ungurahua*. The following list gives the references for these names and it also gives vernacular names from other countries. In parentheses are given the area where the name is used, or the language in which it is used and the literature reference or voucher specimen from which the information was obtained. Names in bold are the ones most commonly used.

Aricaguá, Aricacuá (Colombia, Venezuela, Balick 1986); *Ataíto* (Guahibo, Colombia, Balick 1979); *Axkoé* (Peru, Dahlgren 1936); *Batawa* (Carib, Guyana, Balick 1986); *Batawo* (Carib, Calzada 1980); *Bataua* (Colombia, Balick 1986); *Bocohañu* (Cubeo, Colombia, Balick 1986); *Boreyabeñu* (Cubeo, Colombia, Balick 1986); **Chapil** (Coaiqueres, Ecuador, Barfod and Balslev 1988; Colombia, Balick 1986); *Cohañu* (Cubeo, Colombia, Balick 1986); *Cola-boca, Cola-pa-chi* (Cayapa, Ecuador, Barfod and Balslev 1988); *Cómee* (Colombia, Balick 1986); *Comeé* (Huitoto and Muinane, Colombia, Balick 1986); *Comenyá* (Huitoto and Muinane, Colombia, Balick 1986); *Comenja* (Uitoto, Peru, Dahlgren 1936); *Consa* (Piojé-Tucano, Peru, Dahlgren 1936); *Coroba* (Venezuela, Balick 1986); *Cosa* (Siona, Ecuador, Balslev and Barfod 1987); *Coumou* (?, Jamieson and McKinney 1934); *Cuperi* (Colombia, Balick 1986); *Curuba* (Venezuela, Balick 1986); *Cuumu* (Karijonas, Colombia, García B. 1974); *Cuuruhu* (Bora, Peru, Balick 1986); *Duebocohañu* (Cubeo, Colombia, Balick 1986); *Guacaria* (Makuna, Colombia, García B. 1974); *Guapée* (Matapa, Colombia, García B. 1974); *Hunguravi* (Peru, Venezuela, Balick 1986); *Isá* (Shipibo, Peru, Bodley and Benson 1979); *Itsama* (Chacobo, Bolivia, Boom 1986); *Jagua* (Trinidad, Venezuela, Balick 1986); *Koanani* (Xeriana-teri, Yanomama, Brazil, Anderson 1978); *Kole, Kore* (Uarekena-Arawak, Dahlgren 1936); *Komalhe* (Witoto, Colombia, Balick 1986); *Komboe* (Surinam, Balick 1986); *Kómee* (Muinane, Peru, Dahlgren 1936); *Kuanamré* (Carib, Wayunkomo, Venezuela, Balick 1986); *Kuarámo* (Arawak Baré, Venezuela, Balick 1986); *Kuhéri* (Carib, Maquiritare, Venezuela, Balick 1986); *Kula'po-tci* (Cayapa, Ecuador, Barret 1925); *Kunhua* (Carib, Maquitare, Venezuela, Balick 1986); *Kunúa* (Carib, Pemón, Venezuela, Balick 1986); *Kunwada* (Carib, Kamarakoto, Balick 1986); *Kunyek* (Carib, Pemón, Venezuela, Balick 1986); *Kupéri* (Guahibo, Venezuela, Balick 1986); *Manáca* (Omagua-Cocame, Peru, Dahlgren 1936); *Maripa de Montaña* (Guianas, Balick 1986); *Milpés* (Colombia, Ranghel 1945); *Milpesa, Milpeso, Milpesos* (Colombia, Balick 1986); *Milpesos* (Spanish, Ecuador, Balslev and Barfod 1987); *Mohee* (Warrau, Guyana, Balick 1986); *Muji-ru* (Warao, Venezuela, Braun and Delascio C. 1987); *Ngúndzi* (Mainas, Peru, Dahlgren 1936); *N̄omia* (Tanimucas, Colombia, García B. 1974); *Numuñame* (Guananos, Colombia, García B. 1974); *Oarcéma* (Baré-Arawak, Venezuela, Dahlgren 1936); *Obango* (Colombia, Balick 1986); *Oruta* (Emberá-Chamí, Colombia, Balick 1986); *Osa* (Coto-Tucano, Peru, Dahlgren 1936); *Oxáe* (Guahibo, Colombia, Balick 1979); *Palma de jagua* (Trinidad, Venezuela, Balick 1986); *Palma de leche* (Colombia, Venezuela, Balick 1986); *Palma patavona* (French Guiana, Balick 1986); *Palma real* (Venezuela, Balick 1986); *Palma resina* (Colombia, Balick 1986); *Palma seje* (Venezuela, Cavalcante 1974); *Palma zamora* (Venezuela, Balick 1986); *Patabá* (Yeral, Colombia, García B. 1974); *Patahuá* (Colombia, Balick 1986); *Pataka koemboe* (Surinam, Balick 1986); *Patawa-koemboe* (Surinam, Wessels Boer 1965); *Patauá* (Colombia, Surinam, Balick 1986; Brazil, Anderson 1978); *Patawa* (Colombia, Balick 1986); *Peédi* (Piaróa, Venezuela, Dahlgren 1936); Petowe (Waorani, Ecuador, Wade Davis *et al.* no. 1004) *Pevítsa* (Guahibo, Colombia, Balick 1979); *Pitúma* (Ssabela-Ges, Peru, Dahlgren 1936); *Ponáma* (Chamicura, Peru, Dahlgren 1936); *Punáma* (Arawak, Venezuela, Balick 1986); *Punariá* (Yukunas, Colombia, García B. 1974); *Pupéri* (Tariána-Arawak, Dahlgren 1936); *Sacumana* (Peru, Cavalcante 1974); *Sea* (Peru, Dahlgren 1936); *Segen palm* (English, Ranghel 1945) *Seje, Seje grande, Seje hembra* (Venezuela, Balick 1986, Sirotty and Malagotty 1950); *Shimpi* (Shuar, Ecuador, our obs.); *Shigua* (Quichua, Ecuador, Pedersen no. 67306); *Shinará* (Chayahuitas, Peru, Dahlgren 1936); *Sinami* (Peru, Balick 1986); *Socorrong* (Cholo, Colombia, Balick 1986); *Sokarjo* (Choco, Venezuela, Johnson 1986); *Sucuman* (Peru, Dahlgren 1936); *Thimasé* (Yagua, Peru, Dahlgren 1936); *Tooroo* (Guyana, Balick 1986); *Toru* (Guyana, Balick 1986); *Trupa* (Emberà-Chamí, Colombia, Balick 1986); *Tsá Komak* (Záparo, Peru, Dahlgren 1936); *Tsitsihu* (Bora, Peru, Balick 1986); *Turu* (Arawak, Guyana, Balick 1986); *Uíbn* (Makú, Dahlgren 1936); *Unamá* (Colombia, Venezuela, Balick 1986); *Unamo* (Colombia, Venezuela, Balick 1986); *Unania palm* (English, Ranghel 1945) *Ungurahua* (Quichua, Ecuador, Balslev and Barfod 1987); *Ungurahui* (Peru, Calzada 1980, Kahn 1990); *Ungurauy* (Peru, Balick 1986); *Uruta* (Cholo, Colombia, Balick 1986); *Wógn* (Makú-Nodobo, Dahlgren 1936); *Yacohañu* (Colombia, Balick 1986); *Yagua* (Trinidad, Balick 1986);

Figure 11. A large individual of *Jessenia bataua* at Tena in Amazonian Ecuador. The palm may reach 30 meters height and its leaves may be up to 11 meters long. It is common and widespread in lowland Ecuador and it is well known and appreciated by local people for its oil containing fruits.

Yaro (Arawak Baniva, Guarek, Venezuela, Balick 1986); *Yaveecohañu* (Cubeo, Colombia, Balick 1986).

Morphology

Jessenia bataua is a solitary, large palm with a trunk up to 30 meters tall and a dbh of 20–30 centimeters, occasionally up to 40 centimeters; young trunks are often covered with old leaf sheaths, older trunks are clean with more or less visible nodes (Fig. 11). The leaves are pinnate, erect to spreading, numbering 8–20 on each palm, and measuring up to 11 meters in length; the pinnae number 65–110 per side, and are regularly distributed and borne in one plane. Inflorescence buds are at first interfoliar, later infrafoliar. Inflorescences and infructescences are infrafoliar. *Jessenia bataua* is monoecious. The basal parts of the rachillae have triads of one pistillate and two staminate flowers, the distal ends have staminate flowers in pairs. The short peduncle and the numerous pendant rachillae borne on a short rachis gives the cream colored inflorescences a horse-tail like appearance. The fruits are dark-violet to black when ripe, 3–5 centimeters long and 2–3 centimeters wide with a thin mesocarp and one large seed with ruminate endosperm. The root system is mainly superficial and well developed (Sirotty and Malagotty 1950); adventitious roots spread laterally to a distance of 6–7 meters, but some deeper roots are also found. Association with vesicular-arbuscular mycorrhiza has been observed on *J. bataua* in the greenhouse, but not in the field (St. John 1988).

Distribution

Jessenia bataua is distributed throughout northern South America, in Panama, and in Trinidad (Balick 1986). In Ecuador it is common on both sides of the Andes from sea level up to about 1000 meters (Fig. 12). We have collected it in southern Ecuador along the road from Zamora to Loja where it common at 1100–1200 meters but reaches as high up as 1350 meters *(Pedersen no. 130)*. In Ecuador we have observed *J. bataua* only in humid areas where precipitation exceeds evapotranspiration in every average month. In Brazil it occurs in several areas with a pronounced dry period but these may be swampy areas where ground water is available throughout the year. In localities where *J. bataua* occurs mean annual precipitation varies from 1523–6315 mm and mean annual temperatures from 21.2–27.2 °C. Climatic diagrams 1, 2, 4, 5, 6, 7, 8, B, C, and D shown in Chapter 12 are from stations close to localities where *J. bataua* occurs.

Habitat

Jessenia bataua grows up to 35 m tall and is part of the canopy in areas covered by lowland tropical rainforest, lower montane forest, or gallery forest but it is not known from open, non-forested habitats, probably because it cannot germinate there. Regeneration studies in French Guiana, Venezuela, and Colombia suggest that shade is essential for its germination and early growth, whereas later growth needs light (Braun 1968, CONIF 1980, Sist and Puig 1987). Mortality among juvenile plants is very high,

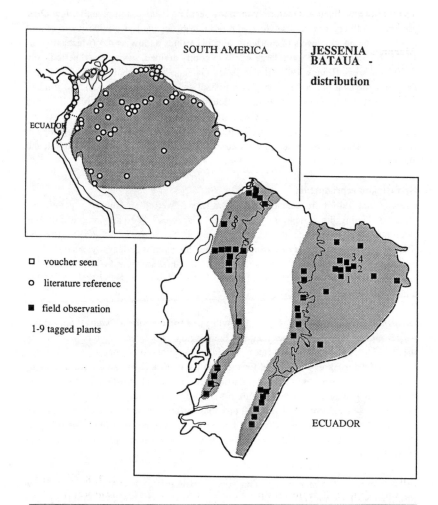

SOUTH AMERICA

JESSENIA
BATAUA -

distribution

☐ voucher seen

○ literature reference

■ field observation

1-9 tagged plants

ECUADOR

Figure 12. Distribution of *Jessenia bataua* in South America and Ecuador. Its total distribution (shaded area) covers the Amazon basin and the lowlands west of the Andes from Ecuador through Colombia to Panama. In Ecuador it is found on both sides of the Andean Cordillera up to 1000 meters above sea level on the slopes. The sign for 'voucher seen' shows the collecting localities for the herbarium specimens cited at the end of this chapter. The sign for 'literature references' shows the location of sites cited in the literature with such exactitude that they could be placed on the map; they are taken mostly from Balick (1986). The sign for 'field observation' show such localities where we observed the palm but did not collect it.

resulting in a population structure with many seedlings and young juvenile, few older juvenile, and very few adults. *Jessenia bataua* grows well in a wide range of soil types, tolerating low levels of nutrients, and pH values as low as 4.3 (Mazzani *et al.* 1975). It is common and forms large stands in poorly drained, swampy, and periodically inundated areas (Balick 1986, Berry 1976, Galeano and Bernal 1987), and in the Peruvian Amazon one hectare on gleyic podsol soils had 130 adults and 50 trunked juveniles of *J. bataua* (Kahn 1988). In Ecuador, where we have observed it almost exclusively on *terra firme,* it grows in small groups or scattered in the forest and Korning and Thomsen (1988) reported 24 trunked individuals in one hectare at Añangu in Amazonian Ecuador where the soil was extremely clayey with high levels of kaolinit and aluminum.

Growth and reproduction

Growth rate, fruiting frequencies, and age at first flowering are important parameters when evaluating the agricultural potential of a species. We encountered only one reference with information about these parameters for *Jessenia bataua* (Sirotty and Malagotty, 1950, states that flowering starts when the palm is about six years old), so in order to obtain information about the range of these values we tagged nine individuals of *J. bataua* at five different localities in Ecuador in April 1987 and checked them in March and April 1988 (Tables 6, 7, 8).

Growth rate and age of trunk. — Growth rates for the nine *J. bataua* individuals vary from 14–72 centimeters per year, and there is no obvious correlation between age of the individuals and their growth rates. Based on extrapolation from one year's growth, the age of the trunk was calculated for the nine palms (Table 6), but the palms may be much older than their trunks since the starting age for vertical growth in *J. bataua* is not known and it may vary because trunkless *J. bataua* probably survive in a waiting position in the understory until a treefall provides sufficient light for vertical growth (Sist and Puig 1987). For comparison, it is known that *Orbignya martiana,* for instance, produces it first divided leaf after seven years and initiates vertical growth after 42 years (May *et al.* 1985).

Inflorescence-bud formation. — Our survey show that each year from 1.1–4.6 new inflorescence buds are exposed, *i.e.* the leaf supporting the bud was lost (Table 6). The lowest inflorescence was found 112 centimeters above the ground, on a trunk with a calculated age of 1.7 years. Such early flowering possibly occurs because the palm was growing in a cleared area, exposed to the sun, and possibly benefitting from nutrients released from the clearing of the natural vegetation. In closed forest we have not observed bud formation below six meters above the ground, but in open areas we have often observed it below three meters. Our observations on bud formation suggest that once *J. bataua* has started to flower, most new leaves will support a bud. Out of 74 leaves, 65 (88%) were found to support inflorescence buds. Of these 65 buds at least 14

Table 6. Growth rates, reproductive age, and reproductive frequency in nine individuals of *Jessenia bataua*. Each palm was marked with a numbered aluminum plate just below the lowest inflorescence or infructescence. The distance from the plate to the lowest leaf was measured. Between the plate and the lowest leaf, number of leaf-scars, inflorescence buds, inflorescences, and infructescences were counted, and notes on their state of development were taken. Trunk height and dbh was measured and number of leaves were counted. The figures are based on measuring intervals from 316 to 373 days, converted to 365 days. Age of trunk and age of trunk at first inflorescence formation was calculated by dividing trunk height with annual growth. Leaf production per year was calculated on the basis of number of leaves at beginning and end of study, and increase in numbers of leaf scars over the study period.

Palm No.	1	2	3	4	5	6	7	8	9
Trunk-height (cm)	420	220	345	1650	1000	1400	1080	960	290
Trunk-dbh (cm)	39	32	32	27	25	33	26	25	29
No. of leaves (beginning)	-	16	14	18	10	12	12	13	13
No. of leaves (end)	-	16	18	20	10	11	12	13	11
Leaf production per year	-	4.6	5.8	6.9	4.0	2.0	4.0	4.0	0
Trunk-growthrate (cm/year)	23	72	14	29	33	29	23	28	39
Trunk-age (year; month)	18;7	3;1	25;0	57;4	30;0	46;7	47;5	34;10	7;6
Trunk-age at 1. infloresc. (year; month)	6;8	1;7	14;9	-	-	19;11	-	-	3;11
1. Inflorescence (height above ground, cm)	150	112	206	-	-	585	-	-	150
New infl. buds per year	1.1	3.5	1.2	4.6	2.0	2.0	3.0	3.0	2.0
Interval between measurements (days)	324	316	318	318	373	373	370	370	370

Table 7. Development observed in the inflorescences of nine individuals of *Jessenia bataua*. Development stages of fruits are illustrated by average fruit lengths, each based on measurements of ten fruits. Figures for small fruits (below 1.5 cm) include calyx. For each palm the inflorescences are listed in the same order as found on the palm *i.e.* with the youngest bud above (no. 1) and increasing age downwards (nos. 2–5).

Palm no.	Period (days)	Inflor. no.	Development (from —> to)
1	324	1	bud —> inflorescence at anthesis
		2	bud —> green fruits (1.2 cm)
		3	bud —> green fruits (3.3 cm)
		4	green fruits (1 cm) —> green fruits (3.4 cm)
		5	green fruits (1.1 cm) —> mature fruits (3.4 cm)
2	316	1	green fruits (1 cm) —> mature fruits (3.5 cm)
		2	mature fruits (3.4 cm) —> old infructesc. without fruits
3	318	1	bud —> green fruits (3.6 cm)
		2	green fruits (1.1 cm) —> green fruits (3.8 cm)
		3	green fruits (2 cm) —> mature fruits (3.8 cm)
4	318	1	bud —> green fruits (1 cm)
		2	bud —> green fruits (2.4 cm)
		3	bud —> purple almost mature fruits (3.4 cm)
		4	green fruits (2.3 cm) —> few old fruits
		5	green fruits (3.2 cm) —> lost infructescence
5	373	1	green fruits (2.2 cm) —> old infructescence without fruits
		2	green fruits (4.5 cm) —> old infructesc. with dry fruits
6	373	1	bud —> lost inflorescence
		2	bud —> green fruits (3.9 cm)
		3	inflorescence (pre anthesis) —> green fruits (4.4 cm)
7	370	1	green fruits (2 cm) —> mature fruits
8	370	1	inflorescence (pre anthesis) —> green fruits (4 cm)
		2	green fruits (1.4 cm) —> mature fruits (4.1 cm)
		3	green fruits (4.1 cm) —> old infructescence without fruits
9	370	4	bud —> green fruits (3.6 cm)

Table 8. Localities for the nine tagged individuals of *Jessenia bataua* for which growth data is shown in Tables 6 and 7. The localities are marked on the distribution map for *Jessenia bataua* (Fig. 12).

No. 1: Prov. Napo, Finca Virgin del Carmen, km 3.5 Coca–Los Aucas rd. Growing as shade tree in coffee plantation. Flat and well drained terrain. Altitude 230 m. Voucher: *Pedersen no. 67309.* Nearest climatic station: Climatic diagram no. 1.

No. 2: Prov. Napo, Coca–Lago Agrio rd. km 10. Pasture with scattered trees and palms. Flat and well drained terrain. Altitude 230 m. Voucher: *Pedersen no. 67306.* Nearest climatic station: Climatic diagram no. 1.

No. 3, 4: Prov. Napo, Estación INIAP-San Carlos, Coco–San Carlos rd. Open pasture with scattered trees. Well drained terrain with small hills. The area cleared in 1977. Altitude: 240 m. Nearest climatic station: Climatic diagram no. 1.

No. 5, 6: Prov. Pichincha 2 km north of Alluriquin. Open pasture with scattered trees in a very hilly terrain. Altitude: 880 m.

No. 7, 8, 9: Prov. Esmeraldas, 10 km north of Quinindé, on the east side of Río Quinindé. Open pasture with scattered trees. Flat and poorly drained terrain. Altitude: 200 m. Nearest climatic-station: Climatic diagram no. 7.

died (22%) before mature fruits were formed. The reason for this is not known, but attacks from *Rhynchophorus palmarum* often destroy or severely damage young inflorescences while they are still enclosed in their bracts (Collazos T. 1987).

Inflorescence development. — At the time of tagging, in April of 1987, palms no. 6 and no. 8 carried emerged inflorescences with flowers that had not reached anthesis. When checked a year later, in April–March of 1988, both these inflorescences had developed large green fruits. In palm no. 1 the one young exposed inflorescence-bud developed into an emerged inflorescence with staminate flowers at anthesis. One infructescence, with small green fruits almost enclosed in the calyx, had developed mature fruits in one year. In palm no. 4 an old inflorescence bud developed an infructescence with almost mature fruits. These observations, combined with the figures from Table 7, suggest that the time needed for the development from a young, newly exposed, inflorescence bud to mature fruits is close to two years; the first year the bud develops and the second year the inflorescence develops into a mature infructescence. Collazos T. (1987) found that development of the inflorescence bud from the time of exposure to the time when the inflorescence emerged varied from 10–18 months, and the time from pollination to mature fruits varied from 10–14 months (based on eight palms).

Figure 13. A specimen of *Jessenia bataua* on the west-Andean slopes near Alluriquin with a recently exposed, cream colored inflorescence, an infructescence with green fruits, and an old infructescence. The time it takes for the recently exposed inflorescence-bud to develop into a mature infructescence is nearly two years (Table 6, 7).

The results obtained in our survey are based on few individuals, a short observation period, and many extrapolations. To obtain more precise data, more individuals, more time, and monthly visits would be needed. In a larger survey it could also be possible to correlate growth and production with soil types, climate, age of the palm, and with different management practices such as application of fertilizer.

Phenology

In Colombia flowering in *Jessenia bataua* takes place over a large part of the year with peaks in periods of low precipitation (Collazos T. 1987). In Surinam flowering occurs during the rainy season from May to August, and fruiting occurs from January to April (Wessels Boer 1965). In Brazil the fruiting season is from September to January (Pesce 1985). In Amazonian Ecuador flowering takes place throughout the year with a peak in February through April which coincides with the end of the dry season that normally occurs from December through March (García S. 1988). During fieldwork in Ecuador in January to May in 1987 and 1988 we observed that all development stages could usually be found in any given area (Fig. 13). It appears that flowering is not strictly correlated with climatic events, and that flowering and fruiting can be limited to seasons or occur more or less throughout the year.

Pollination

Inflorescences of *Jessenia bataua* are protandrous. In Amazonian Ecuador García S. (1988) observed that anthesis is connected with production of scent, and that staminate anthesis lasts three weeks, followed by one week of pistillate anthesis during which the inflorescence temperature increases. Production of nectar has not been recorded and the reward to visitors must be pollen and tissue for consumption and oviposition. One species of *Phyllotrox,* a species of Derelomini, and a species of *Mystrops* are believed to be the main pollinators (García S. 1988) and wind pollination has little or no importance.

Dispersal

Given the widespread use of *J. bataua* fruits, we assume that humans have contributed to its dispersal by bringing fruits back to their villages or taking them along when travelling. Apart from humans, a number of animals are known to consume the fruits; some act as seed dispersers whereas others are seed predators. In French Guiana *Pionus fuscus, Pionites melanocephala, Amazona ochrocephala,* and *Amazona farinosa* may act as short distance dispersers, which consume the mesocarp in nearby trees and leave the seed unharmed, whereas toucans (*Ramphastos tucanus*) and *Penelope marail* possibly disperse the fruits over slightly larger distances (Sist and Puig 1987).

Predation

Among larger animals, squirrels (*Sciurus aestuans*) and peccaries (*Tayassu tajacu* and *T. pecari*) consume the mesocarp and the seeds (Sist and Puig 1987, Kiltie 1981).

Actual and potential uses

Inflorescences. — When young, the inflorescences are edible (Balick 1986). They have a nutty taste but as they grow they soon turn bitter. Indigenous people in Brazil have used the ash from burned young inflorescences as a source of salt (Forero 1983).

Fruits. — The Makunos indians in Colombia regard *J. bataua* as an incarnation of the spirit of female ancestors who still feed the living with the milk of their breasts, symbolized by the *chicha* that is produced from their fruits (Schultes 1974). Indeed, the fruits are worshipped wherever the palm is found, and the oil-rich mesocarp is used for many purposes. Fruits are prepared by soaking them in lukewarm water for about one half hour which softens the mesocarp and makes it easier to peel off the thin and hard epicarp. The *chicha* beverage is prepared from the fruits using the same method as described for *Mauritia flexuosa*. The *chicha* is sometimes boiled down until it contains only a little water and is then used medicinally as a remedy against bloody diarrhoea by the Quichua indians in Amazonian Ecuador (Iglesias 1985).

Oil is extracted from the fruits of *J. bataua* in several different ways. Quichua indians along Río Napo in Amazonian Ecuador boil the fruits in water and leave them for two days. The oil, which then accumulates on the water, is skimmed off (Alarcón G. 1988). Siona-Secoya indians, also from Amazonian Ecuador, make a *chicha* and boil it until only the oil is left (Fig. 14; Vickers and Plowman 1984). To the Guahibo indians in Colombia and Venezuela oil from *J. bataua* and species of *Oenocarpus* played a major role in trade, and a more efficient extraction method was therefore developed (Balick 1986). It involves: 1. Overnight storage of fruits below leaves or plastic, 2. Soaking the fruits in 50°C hot water for some hours, 3. Transferring the fruits to a pot with water near the boiling point, 4. Mashing the fruits for some minutes in a large wooden mortar, 5. Separating the pulp and the seeds with a sieve, 6. Pressing the pulp in a woven press like the one commonly applied when removing liquid from cassava. The resulting liquid is then boiled to remove the water and the oil is ready for use. The same method has been described from Venezuela (Sirotty and Malagotty 1950).

At Las Gaviotas in the Orinoco valley in Colombia a small oil mill has been built, mechanizing the extraction of *J. bataua* oil. The mill was designed and put in operation by the Royal Tropical Institute in Amsterdam on initiative of the directors of the Centro de Desarollo Integrado Las Gaviotas. The equipment was obtained from International Technological Assistance B. V. in Holland. The total cost of installing the oil mill was around US$ 30,000 in 1980 prices (Blaak 1988). According to Blaak a mill can serve an area with 4500–5000 palms, while a minimum of 2300 palms is needed to pay the costs. These figures are based on 30 kilograms as an average annual fruit production per palm. For economic analyses and details on equipment see Blaak (1988).

Figure 14. Fruits of *Jessenia bataua* being prepared for oil extraction by Mrs. Angelina Criollo, wife of the chief of the Siona indians in the Cuyabeno area in Amazonian Ecuador. The fruits are kept in a large bract which covered a young inflorescence before it expanded at anthesis. The fruits are being transferred to a pot in which the oil will be extracted by boiling and skimming off over the cooking fire.

Table 9. Comparison of fatty acid composition in % of *Jessenia bataua* oil, olive oil *(Olea europaea),* and the African oil palm *(Elaeis guineensis).* References: 1. Balick (1986), 2. Eckey (1954), 3. analysis by V. K. S. Shukla, Aarhus Olie (1987).

Fatty acid	lauric	myri-stic	palmi-tic	palmi-toleic	stearic	oleic	lino lenic	lino-leic
	%	%	%	%	%	%	%	%
Jessenia bataua								
mesocarp[3]	6.6	2.5	13.7	0.9	3.0	69.2		1.8
mesocarp[1]			13.2	0.6	3.6	77.7	0.6	2.7
not specified[2]			9.2		5.9	81.4		3.5
Olea europaea								
spanish[2]		0.2	9.5		1.4	81.6		7.0
italian[2]		trace	9.4		2.0	84.5		4.0
Elaeis guineensis								
seeds[2]	46–52	14–17	6.5–9		1–2.5	13–19		0.5–2
mesocarp[2]	trace	1.1–2.5	40–46		3.6–4.7	39–45		7–11

Processing of *J. bataua* oil at Las Gaviotas is somewhat similar to the methods used for *Elaeis guineensis*; the fruits are steamed for about eight hours in kettles fueled by press cake, the mesocarp is loosened from the kernels (which are not damaged) in a diesel engine operated digestor, then the kernels are removed by sifting. To reduce the viscosity and thereby increase the efficiency of the pressing the sifted mesocarp pulp is re-heated and the pressing is then done in a manually operated hydraulic press after which the oil is clarified in tanks and excess water is later expelled by heating (Blaak 1988). The mill has also been used to extract pericarp oil from *Maximiliana maripa,* a species which is also native to eastern Ecuador and there known as *Inayo.*

Jessenia bataua oil resembles olive oil very much, chemically as well as in taste (Table 9). This non-perishable oil is used in many ways: as cooking oil, to preserve meat (Plotkin and Balick 1984), for illumination (Acosta-Solís 1963), and for a number of medicinal purposes. In Amazonian Ecuador we have seen it being used against loss of hair and dandruff. In many areas it is used medicinally against colds, bronchitis, asthma, and tuberculosis (Braun 1968, Balick 1986). In Panama the oil is used as a pain killer (Plotkin and Balick 1984).

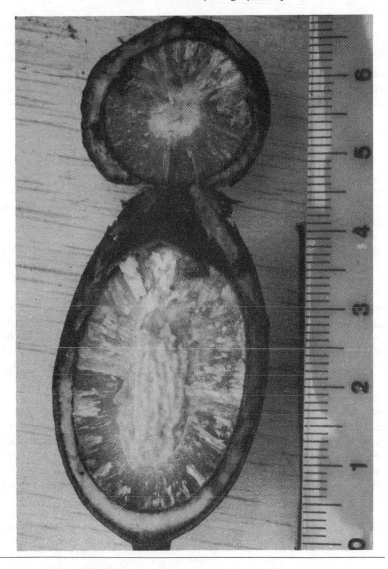

Figure 15. Longitudinal and transversal section of a *Jessenia bataua* fruit. The outer, dark layer is the pericarp (skin), the light homogenous layer, 3–4 mm thick, below this is the mesocarp. In the center the large endosperm is seen; the outer 5 mm show the rumination characteristic of *Jessenia* seeds, a character which *i.a.* distinguishes it from the closely related *Oenocarpus*.

The seeds of *Jessenia bataua* (Fig. 15) are edible, but usually they are not used (Civrieux 1957). One analyses, made by V. K. S. Shukla, Aarhus Olie in 1987, of seeds we collected in Ecuador showed a protein content of 2.9%. Oil content of the seeds is very low, varying from 0.3–3% (Pesce 1985, Markley 1949). Seeds of the closely related *Oenocarpus bacaba* contains various mannites, one galactose, and saccharoses (Moldenke 1949). If large scale oil extraction is initiated the possibility of using the seeds for human consumption, as animal fodder, or for other purposes should be studied.

Though widely appreciated, oil and fruits of *Jessenia bataua* enter the market economy only to a limited extent. In the first half of the century Brazil exported oil from *J. bataua* and species of *Oenocarpus* to USA and Europe. About 1940 the annual export was close to 100 tons. Following the Second World War this export decreased, probably because of increased competition from olive oil and lack of fruits following past destructive harvest (Balick 1985).

In Ecuador *Jessenia* fruits are sold at lowland markets in Coca and Borbón. In 1987 the price was 0.25 US$ per kilogram. Oil, mainly to be used as hair oil, was sold in Puyo in Amazonian Ecuador at 6.5 US$ per liter. In Quito the oil is occasionally sold in health stores.

Trunks. — Edible larvae of *Rhynchophorus palmarum* are collected from decaying trunks. One trunk may yield a kilogram of larvae, each one more than five centimeters long (Balick 1986). The durable wood of the trunk is used for construction and among the Xiriana-teri indians (Yanomama) in northern Brazil for manufacture of bows and arrows (Anderson 1978).

Leaves. — Long stiff fibers from the leaf sheath margins of *J. bataua* are used as blow-gun darts (Wallace 1853) and in Ecuador this same use is recorded from the coastal plain among the Cayapas indians who also use the leaves in initiation rites for young men (Balslev and Barfod 1988). The Waoranis in Amazonian Ecuador use leaf base fibers when cleaning the blow-gun bore and for starting fires (*Wade Davis et al. no. 1004*). Occasionally the leaves are used for thatch. The palm heart is edible, and in Iquitos, Peru, it is sold in restaurants (U. Blicher-Mathiesen, pers. comm.).

Seedlings. — Small seedlings, still attached to the seeds, are used by the Bora indians in Peru as a remedy against snake bites; about 10 seeds are collected, soaked in water for a few minutes, and the liquid is then drunk (Balick 1986).

Roots. — The Waoranis in Amazonian Ecuador use the adventitious roots medicinally in treatment of worms, diarrhoea, headache, and stomach ailments (*Wade Davis et al. no. 1004*).

Management

Despite the widespread use of fruits and oil from *J. bataua* we have found only one record of efforts to cultivate it or to improve natural stands of it. In Colombia secondary forest was enriched with young plants of *J. bataua,* which grew much better there than in open areas (CONIF 1980). Otherwise, management of *J. bataua* appears to be restricted to felling or climbing the wild palms in order to collect its fruits and other products. Management of natural stands of *J. bataua* could involve selective cutting of parts of the vegetation because more light appears to favor its growth and early flowering, but before such cutting, additional *Jessenia*-seedlings should be planted and well established, because shade is essential for their early growth. The cut vegetation could supply nutrients to the soil for a period.

Cultivation. — Germination occurs within 20–90 days (Braun 1968, Jordan 1970, FAO/CATIE 1984) and under adequate conditions germination rates are high. Jordan obtained 99% germination (100 seeds, 83 days) in one experiment and 80% in another (100 seeds, 79 days) using the following method: fruits were harvested when they started to fall from the infructescence, the mesocarp was removed immediately and the seeds were washed in water, to prevent growth of fungi they were dried in the sun for two hours and then placed at a depth of about one centimeter in boxes with sand, the boxes were elevated to provide drainage and a roof provided partial shade, watering was done twice a day.

Germination of *J. bataua* seeds can be stimulated by soaking the fruits in hot water for one half hour, or in warm water (20–25°C) for a week, and removing the mesocarp (FAO/CATIE 1984). Seeds are viable for 3–6 weeks if stored in darkness under not too dry conditions. If sown in a nursery transplanting should be done when the plants are from 1–1.5 years old (CONIF 1980). Whether sown in a nursery or in the field the work by Sist and Puig (1987) suggests that shade should be provided for a period.

Jessenia bataua benefits from association with vesicular-arbuscular mycorrhiza (St. John 1988) and increase in dry weight is almost twice as much in palms with mycorrhiza as in controls without. The beneficial response is believed to be mainly due to more efficient uptake of phosphorus which suggests that it would be worthwhile to secure the presence of mycorrhiza in nurseries, especially if the plants are later transplanted to more open areas where mycorrhiza could be absent because of high soil temperatures. Inoculation of nursery soil with mycorrhiza may be done by mixing the soil with decaying forest litter (St. John 1988) but if seedlings from the wild, rather than from nurseries, are used they may already carry mycorrhiza.

Once planted the palm may start to produce inflorescences in less than two years following initiation of vertical growth (Table 6).

Pests. — Fungi of the genus *Pestallozzia* make large concentric grayish brown spots and may seriously damage the leaves of young *Jessenia*-plants (Sirotty and Malagotty 1950). The beetle *Rhynchophorus palmarum* attacks young palm-inflorescences (Collazos T. 1987) and they can also destroy living trunks, especially if they have been wounded. A species of Derelomini, apart from being a pollinator, lays eggs in 14–63% of the pistillate flowers and the resulting larvae often destroy the ovule (García S. 1988). The Derelomini problem cannot be controlled by pesticides because it would kill other pollinators as well but mechanical control could be possible though costly and work demanding. Such mechanical control could be done by placing, around each inflorescence, a net with holes large enough for penetration by the two small pollinators (*Phyllotrox* sp. and *Mystrops* sp.) but too small for the larger destructive species.

Pests may reduce the reproductive efficiency which, for *Jessenia bataua* when measured as the number of mature fruits per number of pistillate flowers, is from 3–14.5% based on 5 palms (Sist and Puig 1987, Balick 1986). García S. (1988) found that from 60.7–95.5% of the flowers were pollinated (based on 5 palms). Though from different areas these data suggest that pollination failure is not the main reason for low reproductive efficiency. Pests, and maybe abortion due to nutrient deficiency, are more likely reasons.

Felled palms, regardless of the species, attracts ovipositing *Rhynchophorus palmarum*, and palm trunks should therefore not be left to decay in areas of intensive palm-management and cultivation because it will help the beetle to build up large populations.

Production. — During a year a *Jessenia*-palm produces 1.1–4.6 infructescenses (Table 6), each one with 5–25 kilograms of fruits (Balick 1981b, Markley 1949, Sirotty and Malagotty 1950), which gives an annual fruit production per palm of 5.5–115 kilograms. The mesocarp accounts for about 40% of the total fruit weight (Pesce 1985), and oil content of the mesocarp varies from 12.4–18.2% (Pesce 1985) which gives an annual oil yield from 0.3–8.4 kilograms per palm. For comparison, Olive trees *(Olea europaea)* in the Mediterranean area, produce a similar oil and yield an average of 20 kilograms of fruits, corresponding to 1.3–2.6 kilograms of oil per tree per year (Franke 1982). In *Jessenia bataua* there is a considerable, genetically determined, variation in yields, fruit numbers, -size, and -composition that provides abundant material for breeding programs which in one or two generations could probably produce a palm with 50% mesocarp and 25% oil in the mesocarp. Furthermore natural hybrids between *Oenocarpus bacaba* and *Jessenia bataua,* which contain a very small and sterile seed and a relative large mesocarp, could possibly produce more oil than *J. bataua* (Balick 1981b).

Harvest. — With a trunk growth of up to 72 centimeters per year a *Jessenia*-palm will quickly become too tall to be harvested from the ground. Even though its infructescences are fully exposed and easier to reach than those of the African oil palm (*Elaeis guineensis*) they cannot be harvested with the same equipment (a long pole with a curved knife) because the strong and fibrous peduncle of *J. bataua* is too hard to cut. Thus at present there seems to be only two possible methods; climbing or felling the palm.

The loss of nutrients resulting from removal of fruits cannot be calculated because so far *Jessenia*-fruits have been analysed only for protein and oil content. Oil consists of oxygen, hydrogen, and carbon and gives no loss of nutrients from the ground but harvest of proteinaceous products remove nitrogen (N) which in part may be replaced by planting a legume cover crop, for instance *Pueraria* sp. which is used in African oil palm plantations.

Potential in different land-use systems

Jessenia bataua may be more productive than Olive trees and its oil should therefore be able to compete successfully with olive oil if it is accepted by consumers. A major problem regarding commercial exploitation of *J. bataua* is that its fruits deteriorate in a short time which means that scattered sources, whether cultivated or wild, are difficult to exploit. Pesce (1985) suggests that this problem can be solved by harvesting the fruits when they are still immature, and storing them until they can be dried in the sun or by other means.

Extractivism. — Exploitation of wild stands of *Jessenia* for subsistence is important for the nutrition of many people in the Amazon and it can continue to be so if sustainable methods of harvesting are introduced and enforced. In some areas it may be possible to find natural stands of *J. bataua* which are large enough to make the installation of an oil mill worthwhile, but in most areas it would be necessary to increase the density in the stands to make such investments profitable.

Agroforestry. — At present *J. bataua* appears only moderately suited for agroforestry systems, even if it has some beneficial properties: 1. It can be grown on most soil types including acidic and waterlogged soils which are unsuitable for most other crops, 2. The large permanent root system, with a large superficial layer and some deeper roots, helps reduce leaching, and can draw water and nutrients from deeper layers, 3. It supplies good quality oil and protein which can be used by the farmers for consumption and animal ration. Among the disadvantages are: 1. The palm needs shade for initial growth and exposure to light for the production of fruits. Even if these conditions make its cultivation difficult they could be exploited by letting the young establish establish themselves in the shade of, for instance, coffee (*Coffea arabica*) or cacao (*Theobroma cacao*), and later become shade trees for these crops. 2. The extensive superficial root system of *J. bataua* may compete with other crops, especially annuals. 3. Though the palm furnishes many products, most of these are at present of little importance. 4. Given the present lack of oil mills the market for the fruits is very limited.

Plantations. — There are no obvious biologic obstacles to growing *J. bataua* in plantations. In the wild it occurs in large, almost pure, stands but it is not known whether the pests mentioned reduce yields in the large stands to levels significantly below yields from scattered palms. Even fairly small plantations could supply enough fruits for an oil mill, and could make it worth while installing equipment for cracking and milling the seeds so they could be used for animal fodder. A 12 hectare plantation with 200 palms per hectare could supply about 150 tons of fruits per year, which would be enough to justify a mill. These figures are based on the assumption that the production of the palms is doubled by application of 1.5 kilograms of fertilizer per year (Blaak 1988). When the palms grow too tall to be harvested, or the production decreases because of age, the palms must be replaced. A large concentration of palms could may make it economically feasible to can the palm heart and to use the trunks. The durable and beautiful black wood can be used for construction or, like *Cocos nucifera*, for a number of other purposes such as furniture, roof tiles, laminated timber, wood block floors *etc.* (FAO 1986). The rapid growth of *J. bataua* is a problem because it makes harvesting difficult or rotations short. Interplanting with the young shade demanding seedlings before cutting will reduce the nonproductive period.

Conservation status

Jessenia bataua is a widespread species growing in a variety of habitats including swamps that are seldom cleared for agricultural purposes. Thus, even if it has been seriously depleted in many areas due to destructive harvest, and even if it lacks the ability to regenerate in the ever increasing open areas there are no immediate threats to the palm at the species level. A crucial question right now, however, is whether its ability to grow in many different habitats is due to a large number of genetically different ecotypes, or whether any individual possesses this wide adaptation. Transplanting experiments could provide the answer, but at present it is not known. If a large number of ecotypes exist, then many of them are endangered. One example is the high altitude stands found near Loja in Ecuador which, if they constitute an ecotype, will very soon be endangered due to forest clearance in the area.

Within populations selective, destructive harvest may lead to loss of genetic variation. In Venezuela, due to cutting rather than climbing, the strongest, largest, and most productive palms are seriously depleted every year (Sirotty and Malagotty 1950). The International Union for Conservation of Nature (IUCN) lists the status of *Jessenia bataua* ssp. *bataua* as not threatened and that of *J. bataua* ssp. *oligocarpa* to be indeterminate in Trinidad and unknown in the rest of its distribution area (Dransfield *et al.* 1988).

Research

Jessenia bataua, along with *Bactris gasipaes,* has received much attention, compared to other native palms of the Neotropics, and several research institutions work on domestication programs with these two palm species. Most of the work on *J. bataua* is coordinated by Programa Interciencia de Recoursos Biológicos (PIRB). In Brazil experimental plantations of *Jessenia bataua* have been established by CEPELAC outside Itabuna and at CEPATU in Belém. The latter is a germplasm bank coordinated by EMBRAPA and CENARGEN (Balick 1988). Michael Balick of The New York Botanical Garden carries out extensive work on the systematics and economic botany of the *Jessenia-Oenocarpus* complex (Balick 1981b, 1986, 1988).

Specimens studied

The following herbarium specimens were seen or collected for this study. For each specimen the following information is given: *Collector and number (in italics)*, locality, vegetation, elevation above sea level, date, and the acronyms of the herbaria where the specimens are deposited.

Pedersen, H. B. no. 67306. Ecuador, prov. Napo. Coca–Lago Agrio rd. km 5 (76°59'W; 00°26'S). Pastures with open vegetation. 255 m. 7 Mar 1988 (AAU, QCA, QCNE); *Pedersen, H. B. no. 67309.* Ecuador, prov. Napo. Finca Virgin del Carmen. Coca–Los Aucas road km 3.5 (76°59'W; 00°29'S). Palm left in coffee plantation. 250 m. 7 Mar 1988 (AAU); *Pedersen, H. B. and Madsen, J. E. no. 130.* Ecuador, prov. Zamora-Chinchipe, Loja–Zamora rd. 1340 m. 30 May 1989 (QCA); *Wade Davis, E., Yost, J. and Tomo no. 1004.* Ecuador, prov. Napo. Confluence of Quiwado and Tiwaeno Rivers (77°40'W; 1°50'S.) 20 Apr 1981 (QCA).

10. AMMANDRA NATALIA

Ammandra natalia is a conspicuous and important element in the vegetation around its type locality in the province of Morona-Santiago in Amazonian Ecuador where it also is of great economic importance because its fibers are used for the manufacture of brooms throughout Ecuador. The fibers are harvested mainly from wild palms, but cultivation has begun in some areas, and, apart from fibers, *A. natalia* produces edible fruits, edible palm heart, and its leaves are used for thatch. This multipurpose palm seems ideally suited for small scale farming systems, and used in pastures it increases the economic yield several fold at the same time as it increases the stability of this simple silvo-pastoral system.

Taxonomy
The genus *Ammandra* with only two species, *A. decasperma* and *A. natalia*, belongs to the palm subfamily Phytelephantoideae which was recently revised by Barfod (1988). *Ammandra decasperma* is distributed west of the Andes on the Pacific coastal plain of Colombia. *Ammandra. natalia* grows east of the Andes and was described from material collected near Sucua in southern Ecuador by Balslev and Henderson (1987a); it is so distinct that Barfod considers erecting a new monotypic genus for it.

Vernacular names
The following list of vernacular names gives the references for these names. In parentheses are given the area where the name is used or the language in which it is used and the reference or voucher specimen from where the information was obtained. Names in bold are the ones used most commonly.

Chilli (Quichua, Ecuador, Pedersen no. 67305); *Chilli-punschu* (Quichua, Ecuador, Pedersen no. 67305); **Fibra** (Spanish, Ecuador, Pedersen no. 67305); *Pissaba* (Spanish, Ecuador, Balslev *et al.* no. 62465); *Sili* (Quichua, Ecuador, Orr and Wrisley 1981); **Tindiuqui** (Shuar, Ecuador, Pedersen no. 67305); *Wamowe* (Waorani, Ecuador, Wade Davis and Yost no. 997).

Morphology
Ammandra natalia is a solitary, medium sized palm with a trunk that is up to five meters tall with a dbh up to 30 centimeters, and it is completely or partly covered with persistent leaf bases (Fig. 16). The pinnate leaves are erect to spreading, numbering from 10–36, and measuring up to five meters in length; their bases and petioles split up into a conspicuous fibrous cover, and the regularly distributed, 90–100 pinnae per side are borne in a horizontal plane except for those in the apical part, which are borne vertically because the rachis twists. The palm is dioecious, the inflorescences are interfoliar, and the infructescences are inter- or infra-foliar. The dense, long, and pendant fleshy staminate inflorescence may reach a length of more than two meters which, combined with its light yellow cream color, makes it very conspicuous. Almost globular, brown, and

Figure 16. *Ammandra natalia* growing near Sucua in Amazonian Ecuador. The male inflorescences, up to two meters long, hang out from the leaf axils. The lower leaves have been cut off and the fibers from their bases have been harvested. The palms grow in a cattle pasture with ground cover of the grass *Axonopus scopamus*.

dense clustered infructescences are formed from the erect and dense female inflorescences, each cluster having a diameter of about 30 centimeters and containing 30–40 fruits. From a distance it looks as if the clusters, of which five or more may be found on a single palm, are sessile on the trunk.

We excavated the root system of a free growing palm with a trunk 4.45 meters tall and a dbh of 27 centimeters, growing at 635 meters above sea level in a pasture on stony, well drained *terra firme*. The root system was superficial and above the ground we found a dense cone of adventitious roots, 35 centimeters tall and 55 centimeters in diameter at the soil surface, which continued below the ground to a depth of 40 centimeters, reaching a diameter of one meter. Continuing in the lateral and horizontal directions the cone quickly disintegrated; in a lateral distance of 60 centimeters from the trunk very few roots were found below 50 centimeters and none were found below 60 centimeters. The root system extended to a distance of 4.5 meters from the trunk at a depth of 30 centimeters, but beyond 2.2 meters from the trunk roots were few, and found only from 15–30 centimeters below ground. The roots had a diameter up to eight milimeters, and were black with lenticels and only few side-branches.

Distribution

Ammandra natalia is widespread in Amazonian Ecuador and Alwyn H. Gentry from the Missouri Botanical Garden told us that he has seen the palm, and the use of its fibers for making brooms, in Amazonian Peru (Fig. 17). It grows spontaneously up to elevations of 800 meters above sea level and is cultivated up to 1000 meters near Sucua in Amazonian Ecuador (Climatic diagram no. 3, Chapter 12). Where *A. natalia* occurs the climate is warm and humid throughout the year and the average precipitation exceeds potential evapotranspiration in every average month, annual mean precipitation varies from 1664–6315 mm, with no month having a mean below 94 mm, and annual mean temperatures varies from 21.7–25.3°C. Climatic diagrams no. 3, 4, 5, and 6 in Chapter 12 are from stations close to localities where *A. natalia* occurs.

Habitat

Ammandra natalia occurs spontaneously, in areas of *terra firme* covered by evergreen tropical rainforest, as a slow growing sub-canopy palm that reaches a maximum height of about 10 meters including the leaves, but it also seems to thrive and reproduce well in open areas where the forest has been cleared. The leaves have thick adaxial cuticula, small stomata without substomatal chambers, and vascular tissue surrounded by thick sclerenchyma layers (Barfod 1988), features which usually are interpreted as adaptation to periodic drought, which is always more severe in open areas than within the closed forest. This apparently does not agree with its present ecological preferences, but it could suggest that the palm evolved during drier climatic conditions that those that dominate Amazonian Ecuador at present. The density in natural forest may be high and in a 0.1 hectare plot in a deforested area we counted 19 adult *Ammandra natalia*.

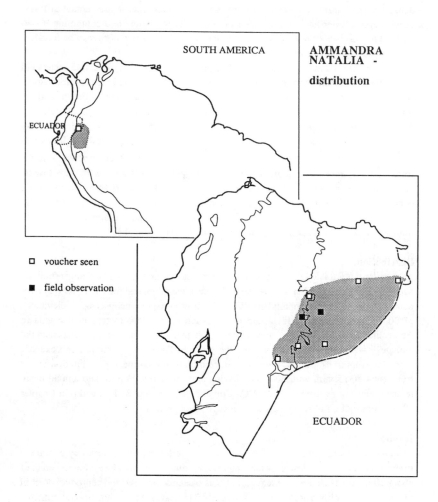

Figure 17. Distribution of *Ammandra natalia* in South America and Ecuador. Its total distrubution (shaded area) is limited to the Amazonas of Ecuador and norther Peru. The sign for 'voucher seen' shows the collecting localities for the herbarium specimens cited at the end of this chapter. The sign for 'field observation' show such localities where we observed the palm but did not collect it.

Phenology
No distinct flowering or fruiting season is known, but according to Mr. Carreno, who exploits the palm south of Sucua in Amazonian Ecuador, the first flowering occur about five years after germination which is before a visible trunk has been formed.

Pollination
Pistillate and staminate buds heat up before anthesis, both are scented at anthesis, and no nectar is produced, all of which are factors often associated with beetle pollination (Balslev and Henderson 1987a, Henderson 1986). We collected the insect visitors on a staminate inflorescence at anthesis and these were identified and commented on by our colleague J. M. Olesen; four species of Staphylinidae, one species of Nitidulidae, one species of Chrysomelidae, two species of *Cyclocephala* (all Coleoptera), one species of Drosophilidae, and one species of Phoridae (Diptera) were found inside the inflorescence. One species of Staphylinidae was abundant, and all other species were represented with one or a few individuals. We observed two species of Trigonidae and four species of Apidae (Hymenoptera) flying to and from the inflorescence. With the exception of the Staphylinidae the species of Coleoptera and Hymenoptera may be pollen eaters/collectors and the species of Diptera may be mating and ovipositing in the inflorescence. Some nitidulid beetles (*Mystrops* spp.) are well known pollinators in palms (Henderson 1986), but it can not be ascertained whether the one found here pollinates *A. natalia*.

Dispersal
According to residents in the Sucua area, fruits of *A. natalia* are consumed and dispersed by a small rodent ("guatusa", *Dasyprocta* sp.). Since it has edible fruits and furnishes several other products dispersal by humans has probably also contributed to its present distribution.

Predation
Rodents of the genus *Dasyprocta* may damage the seeds, but otherwise no predators are known. The dull and rigid fibrous warts that cover its fruits may offer mechanical protection as well as camouflage (Uhl and Dransfield 1987).

Actual and potential uses
Inflorescences. — The staminate inflorescences are often eaten by cattle, either directly from the palms or harvested and fed to them. This use of the inflorescence is surprising because they have plenty of of raphide bundles in their cells, which with examples from other plant species are known to cause internal bleedings in humans and cattle if they are ingested (Barfod 1988). This matter should be studied further before any intensified use of the inflorescences for fodder can be recommended.

Fruits. — The oily, orange mesocarp of the fruits (Fig. 18a) is consumed by *colonos* and indigenous people, and in 1987 and 1988 we observed that fruits were sold on the

Figure 18. a. Fruits of *Ammandra natalia* showing the warty, hard exocarp, the soft mesocarp, and the hard seeds. **b.** Cleaning of *Ammandra natalia* fibers on a card made from a peace of wood with nails. The gentleman in the foreground is Mr. M. Carreno who is a fiber farmer in the Sucua area of Amazonian Ecuador.

market in Sucua in Amazonian Ecuador for 20 sucres each (1 US$ = 270 sucres) which appears to be a high price considering the abundance of the palm in the area. Apart from being liked by humans, the fruits attract many rodents, which are often hunted near the palm trees, and they are a most welcome supplement to the diet of many people.

The fluid endosperm from immature fruits is consumed as a beverage. The mature endosperm is very hard as in other species of Phytelephantoid palms, but the seeds are not used. They are smaller, but could be used as a source of vegetable ivory and animal fodder in the same way as those of *Phytelephas aequatorialis* (Chapter 3).

Trunks. — The Waorani indians of Amazonian Ecuador make ceremonial head bands and darts from the trunk of *Ammandra natalia* (*Davis and Yost no. 997*). Edible larvae, probably of the beetle *Rhynchophorus palmarum* which is known to deposit eggs in trunks of many palm species, are collected from decaying trunks. In Sucua the larva is called *Muquindi*, and, besides being a delicacy, it is also considered a pest because it destroy the palm.

Leaves. — Leaves of *Ammandra natalia* are used for a variety of purposes. The palm heart is consumed as a salad. Young leaves are used by lowland Quichua indians as sheaths for protection of poisoned blow-gun darts when hunting according to Mr. B. Bormann, who has lived in Amazonian Ecuador for many years. In the Quichua village of Canelos in Amazonian Ecuador, the leaf-rachises are used as beams for a thatch made of *Carludovica palmata* leaves. For this purpose the rachises are split longitudinally with a *machete,* the parts are tied to the roof rafters with a distance of 25–30 centimeters, the rachis is placed horizontally with the pinnae pointing downwards, and upon these the leaves of *C. palmata* are fastened by wrapping the petiole around the split leaf-rachis of *A. natalia*. A thatch made in this manner will last for about three years. We also observed thatch made entirely of *A. natalia* leaves in Canelos. Waorani indians use *Ammandra* leaves for baskets and leaf sheath- and petiole-fibers are used for brooms, fire starters, and torches (*Davis and Yost no. 997*).

The fibers from the leaf-base of *Ammandra natalia* resemble those harvested from *Attalea funifera* and *Leopoldinia piassaba* in other parts of South America and they are much in demand in Ecuador for brooms. They are harvested throughout the year but the fibers from a given palm are harvested only once a year. After harvest the fibers are cleaned with a toothed instrument, similar to a fakir pillow, made from a piece of wood with nails hammered through and used more or less like a wool cleaning card (Fig. 18b). When clean the fibers are cut into proper length with a *machete*, bunched, and are then ready for sale. For their further processing into brooms small bunches of fiber are bent and fastened with a metal staple in each of the holes on a brooms head.

Large fiber-producers sell their product directly to merchants from the Andean region and the coastal plain, whereas small producers and gatherers sell to local merchants, who, in turn, re-sell them. Part of the fiber is processed into brooms in Nuevo Rocafuerte, Sucua, Canelos, and Puyo, and other Amazonian towns, but most is

Figure 19. Harvest of *Ammandra natalia* fibers near Sucua in Amazonian Ecuador.
The cut leaves are being used as platforms to stand on for easier acces to the remaining
leaves. A minimum of six leaves should be left in order to secure the continued growth of
the palm. The palm stands in a pasture with ground cover of the grass *Axonopus
scopamus*.

exported to broom factories in Cuenca, Quito, Sto. Domingo, and Guayaquil. The price in January, 1988, was about 170 sucres per kilogram in Sucua and 220 sucres per kilogram in Puyo, and working men in Sucua, who could harvest and clean about 20 kilograms of fibers in a day, were paid 300 sucres per day plus meals. One kilogram of fiber is enough for six brooms, selling at 160 sucres each (1 US$ = 270 sucres).

The demand for fibers exceeds the supply, so fiber-producers and -gatherers are always able to sell their products, while broom producers periodically report lack of supplies. The difficult transport situation from the interior influence the supply situation heavily because the high cost of transportation leaves very little to the fiber gatherers who may not find it worthwhile to collect the fibers. Airplanes returning to Puyo from remote villages often bring fiber.

Management

Cultivation. — Cultivation of *Ammandra natalia* appears to be increasing in the southern part of Amazonian Ecuador, but so far most fibers are harvested from wild palms which involves only limited management. This limited management usually involves favoring *A. natalia* by removing other vegetation and by leaving palm-leaves, -infructescences, and -inflorescences, which are cut when harvesting the fibers, on the ground where they supply organic matter, increase the nutrient retention capacity of the soil, and supply extra potassium (Hecht 1983). We have not observed application of fertilizers or pesticides, though the palms do receive some manure because the cattle often seek shade below them.

There is a considerable variation in fiber-yield between individuals, but we are not aware that any breeding or selection has been undertaken to increase the yield. Selection and breeding work would involve collecting of seeds or seedlings from high yielding individuals or stands and artificial pollination with pollen from other high yielding individuals. Artificial pollination should be easy to carry out because the palm is dioecious, the staminate inflorescences contain large amounts of pollen and are easy to spot and reach, the palms are low and the dense structure of the pistillate inflorescence makes it easy to apply pollen at the right place, and it is possible to exclude spontaneous pollinators by covering the inflorescence with a bag. In addition the scent emitted from the inflorescence at anthesis makes it easy to determine the right time for pollination.

To study the effect of different management practices, we established two study plots, each covering 0.1 hectare, five kilometers south of Huambi in the Province of Morona-Santiago (78°10'W; 2°35'S), at 635 meters above sea level in a flat and well drained terrain. Plot no. 1 was established in an area used exclusively for fiber production in which almost all vegetation, apart from the palms, had been cleared, the weeds were kept down, and grazing was excluded. Plot no. 2 was established nearby in an area, which, except for the palms, was cleared 20 years ago and is now covered with *Axonopus scopamus* grass, and used for grazing (Table 10).

The density of palms where grazing is excluded (plot 1) corresponds to 560 fiber producing individuals (size category 1) in one hectare (56 in 0.1 ha.), increasing during the next two years with 390 individuals (size category 2) to 950 individuals in production in one hectare at which point some thinning may be necessary, because the quality of the fibers is not optimal when the palms grow under shaded conditions. Stands where grazing is excluded produce large numbers of seedlings and may serve as nurseries.

The density of palms where grazing is not excluded (plot 2) corresponds to 190 individuals per hectare (19 in 0.1 ha.) and this, the owner said, was the highest number possible without negatively affecting the grazing. There was no obvious competition between the grass and the palm, but we were told that in dry periods the palms would dry out the soil and limit the growth of the grass. *Axonopus scopamus*, however, is adapted to wet soils, and the problem might be solved by growing other species of grass. Competition for light is low because the leaves are erect and frequent cutting of leaves for fiber reduces the crown size; shade covered a circular area up to four meters from the trunk before harvest and after harvest the shade reached only two meters from the trunk. Competition for nutrients is probably small because only fibers are removed, and the residues are left to rot in the field. The lack of regeneration in plot no. 2 could be due to frequent cutting of infructescences, which leaves very few seeds; destruction of seedlings by cattle also contribute to the lack of regeneration.

Pests. — Beetles, probably *Rhynchophorus palmarum*, are a serious pest that often kill large numbers of palms. If the trunk is wounded or the leaves are cut too close to the trunk during harvest this will provide access for the beetles and in order to reduce this problem the basal part of the petiole should be left when harvesting.

Harvest. — Fibers and fruits may be harvested before a visible trunk has been formed (which happens five years after germination) and both are easy to harvest because *A. natalia* never grows very tall. It has been a common practice in some areas to fell the palms when harvesting fibers. Usually, however, the leaves are cut at the base of the leaf sheath, using a *machete* and when the leaves have been removed the fibers can be ripped off by hand. The harvest is done from the ground or with the use of ladders and rope and often partly cut off leaves are used as a working platform (Fig. 19). According to Mr. H. Carreno, who has harvested fibers for twenty years, a minimum of six leaves must be left to secure the continued growth and survival of the palm.

Production. — One palm may yield 4.5 kilograms of fibers in one year, but the individual variation is large, and most palms yields less according to Mr. Carreno. Fruit production is not known.

Table 10. Numbers of individuals in four size categories of *Ammandra natalia* in two 0.1 hectare plots five kilometers south of Huambi in the Province of Morona-Santiago (78°10'W; 2°35'S), at 635 meters above sea level in a flat and well drained terrain. Plot no. 1 with all other vegetation, except the palms, cleared; plot no. 2 in a pasture cleared 20 years ago. Size categories: 1. Palms that had been harvested one or more times. 2. Palms, not previously harvested, with leaves more than three meters long (will be in production within 0–2 years). 3. Palms, not previously harvested, with leaves from 1–3 meters. 4. Palms, not previously harvested, with leaves less than one meter. Size category 4, was not counted in plot no. 1, but seedlings were numerous, many covered by harvested leaves.

	Numbers of individuals	
Size category	Plot no. 1	Plot no. 2
1 (harvested ≥ 1 time)	56	19
2 (not harvested, will produce in 0–2 yrs.)	39	0
3 (not harvested, leaves 1–3 m)	29	0
4 (not harvested, leaves < 1 m)	-	not counted
Total	124	19

Potential in different land-use systems

The main product from *A. natalia*, the fibers, can be stored for a long period without deteriorating and the same is true of the seeds, which may be commercialized in the future. As previously discussed this is of vital importance to small scale extractivism and cultivation in remote areas. An other important aspect is that the fibers can be harvested at any time of the year and the harvest does therefore not interfere with other more seasonal agricultural activities.

Extractivism. — The actual situation demonstrates that exploitation of wild stands of this palm is economically feasible and may be carried out in a sustainable way. Selective removal of some of the other vegetation in areas were the palm is common is an easy and inexpensive type of management, which increases the density of the palm significantly and the removal of the nutrient poor fibers is not likely to change the nutrient status of the soils.

Agroforestry. — *Ammandra natalia* appears to be ideally suited for small scale farming systems, some of the advantages having been mentioned above. Other advantages are:

1. Harvest of fibers remove only few nutrients.
2. The palm produces very little shade because of the continuous cutting of leaves and

intraspecific competition for light is therefore not too severe depending on the choice of crops. The cut-off leaves, which are left to rot, will increase the amount of organic material in the soil.

3. Because *Ammandra* tolerates high as well as low light intensities it can grow in shade or partial shade of other trees, or it can grow with the crown in the upper layer of an agroforestry system, fully exposed to the sun.

4. Because it is a tree crop the palm increases the stability of a system that may otherwise be dominated by annuals, and its permanent root system will act as a nutrient trap, reducing leaching and facilitating nutrient recycling. However, the shallow root system cannot draw nutrients from deep layers.

5. *Ammandra natalia* is a multipurpose palm, that supplies a number of different products. Therefore, when the price of fibers is too low to make harvesting worthwhile, the palm can serve other purposes.

6. Very small investments are needed for the management of this crop plant. A *machete*, a rope, a ladder, and the card described above is enough.

7. The marketing of the product is already organized and the fibers are at present easy to sell.

Used in pastures, whether on small or large cattle farms, the palm is a valuable component with the mentioned advantages. Mr. H. Carreno, the owner of the area where study plot no. 2 was made, earns several times more on fibers than on cattle and palms and cattle occupy the same land with little competition.

Plantations. — As shown in plot no. 1, *Ammandra. natalia* may grow with very high densities under plantation-like conditions. This may favor pests not known at present, but apart from this, there are no obvious obstacles. Many of the above mentioned advantages apply for the plantation situation as well.

Conservation status

Ammandra natalia is distributed only in Amazonian Ecuador and Peru, and it is therefore not a widespread species. In some areas, for instance around Puyo and Canelos, destructive harvest and clearing of land for pastures and agriculture has drastically depleted populations of it but many factors will help to secure large parts of the genetic variation of this species; the large Yasuni National Park lies within its distribution area, cultivation of it has begun and appears to be increasing, the palm seems to thrive and regenerate in open areas, the apparent adaptation to open habitats suggests that it will survive deforestation better than most other palms. The situation for this species is therefore not alarming, even if the danger exist that increased destructive harvest, complete clearing of land for pasture and crops, and the reduced reproduction resulting from cutting inflorescences and infructescences during harvest may alter the present situation. In our opinion *Ammandra natalia* may be assigned the IUCN category 'not threatened'.

Specimens studied

The following herbarium specimens were seen or collected for this study. For each specimen the following information is given: *Collector and number (in italics)*, locality, vegetation, elevation above sea level, date, and the acronyms of the herbaria where the specimens are deposited.

Alarcón, R. no. 123. Ecuador, prov. Napo. Nuevo Rocafuerte y la orilla del Río Napo hasta 5 km al Oeste, y la orilla del Río Yasuni, hasta la Laguna de Jatuncocha (QCA); *Balslev, H., Henderson, A. and Kristensen, F. B. no. 62080.* Ecuador, prov. Napo. 65 km north of Puyo along the road to Tena (77°48'W; 01°07'S). Cut over forest. 550 m. 3 May 1986 (AAU); *Balslev, H., Henderson, A. and Kristensen, F. B. no. 62205.* Ecuador, prov. Morona-Santiago. One hours walk west of Taisha along trail to Cangaime. (77°30'W; 02°25'S). 400 m. 13 May 1986 (AAU); *Balslev, H., Bergmann, B., and Pedersen, H. B. no. 62465.* Ecuador, prov. Napo. Río Napo, ca 5 km downriver from Pañacocha (76°03'W; 00°27'S). 16 Apr 1987 (AAU); *Balslev, H. and Henderson, A. no. 60651* **Type**. Ecuador, prov. Morona-Santiago. Road from Mendez to Sucua km 18, just south of Logroño (78°11'W; 02°35'S). Pastures with leftover trees in area of premontane forest, ca. 800 m. 14 Jul 1985 (holotype AAU; isotypes BH, K, NY, QCA, QCNE); *Davis, W. E. and Yost, J. no. 997.* Ecuador, prov. Napo. Confluence of Quivado and Tiwaeno rivers. 18 Apr 1981 (F, NY, QCA); *Pedersen, H. B. no. 67305.* Ecuador, prov. Morona-Santiago. Sucua–Mendez road, 5 km south of Huambi. (78°10' W; 2°35' S). Flat and well drained terrain, 690 m. 27 Feb 1988 (AAU, QCA, QCNE).

11. EUTERPE CHAUNOSTACHYS

Euterpe chaunostachys is distributed in a limited area on the Pacific coast of Colombia and Ecuador where it is common in swampy areas in river estuaries. Its fruits and palm heart have for many years been harvested by indigenous people and *colonos,* and today large natural stands of it constitute the basis for commercial palm heart industries, and serious over exploitation has been the result even if this palm is ideally suited for extractivism. It is also well suited for small scale farming systems; it provides a number of different products, and it is easy to grow and harvest. The habitat of the palm is threatened by shrimp farming and cultivation of rice (*Oryza sativa*).

Taxonomy

Euterpe belongs to the palm subfamily Arecoideae, and comprises about 28 species (Uhl and Dransfield 1987). The circumscriptions of many *Euterpe*-species are dubious and a taxonomic revision is badly needed. In the present work *Euterpe cuatrecasana* from southwestern Colombia is considered conspecific with *Euterpe chaunostachys* from northwestern Ecuador. They grow in the same habitats, within the same general area, and no significant differences have been encountered, neither in descriptions nor in collected material. On the basis of priority of publication the name *E. chaunostachys* is used but it is possible that a taxonomic revision will show that, what we here call, *E. chaunostachys* is the same species as *Euterpe oleracea* from the Amazon (Galeano and Bernal 1987). Therefore information about *E. oleracea* is also included here, though always under that name.

Euterpe chaunostachys was described by Burret (1929c) and the original description was based on a collection by Eggers (no. 15669) from El Recreo, near Bahia de Caráquez on the coast of Ecuador.

Vernacular names

Three vernacular names for *Euterpe chaunostachys* are known from Ecuador; *Mamba-san-chi* is used by the Cayapa indians (Barfod and Balslev 1988), Eggers (no. 15669) gives the name *Bambil,* and *Palmiche* is a common Spanish name from which the name *Palmichales* for stands of *E. chaunostachys* is derived. In Colombia the common name *Naidi* has given name to stands, called *Naidisales* (Patiño 1977). The most widespread vernacular name for *E. oleracea* in Brazil is *Açaí* (Anderson 1988). The following list gives the references for these names and it also gives vernacular names from other countries. In parentheses are given the area where the name is used or the language in which it is used, and the reference from where the information was obtained. Names in bold are the ones most commonly used.

Bambil (Spanish, Ecuador, Eggers no. 15669); *Mamba-san-chi* (Cayapa, Ecuador, Barfod and Balslev 1988); *Murrapo* (Saija, Colombia, Patiño 1977); **Naidi** (Colombia, Patiño 1977); *Naidisa* (Noanama, Colombia, Patiño 1977); *Nandisa* (Tapaje, Colombia, Patiño 1977); **Palmiche** (Spanish, Ecuador, Pedersen no. 67301).

Figure 20. a. A natural stand of *Euterpe chaunostachys* near Borbón on the Pacific coast of Ecuador. **b.** Harvesting of *E. chaunostachys* for palm heart involves felling of about half of the trunks over seven centimeters diameter in natural stands, here near Borbón (Table 12).

Morphology

Euterpe chaunostachys is a gracious, medium sized cluster palm, often with more than 10 mature trunks in one cluster, each trunk measuring up to 15 meters or more and with a dbh of 5–15 centimeters (Fig. 20a). The palms often grow on small mounds created by the adventitous roots. The leaves are pinnate, curved, and spreading to almost pendant, numbering around 10 and measuring up to four meters in length; the leaf sheath encircles the trunk for about one meter, forming a glabrous reddish brown crown-shaft; the pinnae are regularly distributed, two ranked, more or less pendant, and number 40–80 per side. The inflorescences are infrafoliar, erect, and cream colored with rachillae that point straight out and become more or less pendant as the fruits grow. The fruits are sub-globular, up to two centimeters in diameter, almost black, with one seed and the endosperm is ruminate. The seedlings have bifid leaves. Lenticels and pneumatophores are present on the roots (Acosta-Solís 1971, Lopez O. 1982). The root system of *E. oleracea* is superficial (Granville 1974) and VA-mycorrhiza is associated with the roots on cultivated and wild *E. oleracea* (St. John 1988).

Distribution

Euterpe chaunostachys is found in a narrow belt along the Pacific coast of southern Colombia and northern Ecuador where it occurs up to elevations of 200 meters above sea level (*Bernal and Galeano no. 670*), but it is more frequent near sea level (Fig. 21). In localities where *E. chaunostachys* occurs mean annual temperatures varies from 25.4–27.6°C and mean annual precipitation varies from 2128–7089 mm. Climatic diagrams 8 and G in Chapter 12 are from stations close to localities where *E. chaunostachys* has been collected. Because many of the localities are subject to daily flooding, precipitation is probably of minor importance. *Euterpe oleracea* is very common in the Amazon including Amazonian Ecuador, and also in the Amazon estuary where a dry season occurs (Climatic diagram D in Chapter 12).

Habitat

Euterpe chaunostachys is common in low lying areas and in estuaries along the coast where it grows scattered, in groups, or in large dense stands. We made a 0.1 hectare study plot in a stand near Borbón which had 37 clusters, and a total of 126 trunks with a dbh over seven centimeters. In Colombia surveys based on a total of 31.3 hectares showed an average of 605 clusters and 2758 trunks with dbh over three centimeters per hectare (Lopez O. 1982). The stands are often found in flat areas which are regularly inundated by the tidal back up and the soils in these areas are mainly made up of organic material and deposits from the inundations by the river for which reason they have high levels of nutrients, but low decomposition rate because of permanent water-logging. *Euterpe chaunostachys* has some tolerance to salt (Acosta-Solís 1971, Lamb 1959, Patiño 1977) and in Colombia it is found in associations called *natal* which are dominated by *nato*, a species of *Mora* just above the upper limit of the mangrove forest, and occasionally exposed to brackish water (Lamb 1959).

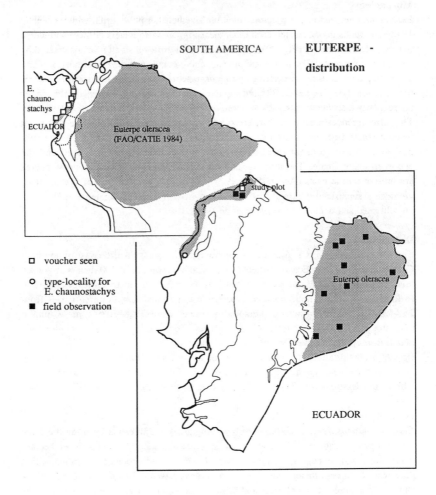

Figure 21. Distribution of *Euterpe chaunostachys* and *E. oleracea* in South America and Ecuador. The total distribution (shaded area) of the two species covers the Amazon basin and the lowlands west of the Andes from Ecuador through Colombia. In Ecuador the populations on the Pacific coast are here treated as *E. chaunostachys*, whereas those in Amazonia are treated as *E. oleraceae*. The sign for 'voucher seen' shows the collecting localities for the herbarium specimens of *E. chaunostachys* cited at the end of this chapter. The sign for 'field observation' show such localities where we observed the palm but did not collect it.

Euterpe chaunostachys appears to be important in the vegetation succession of periodically inundated areas; it is one of the first species in a serial succession (Galeano and Bernal 1987), and Lopez O. (1982) described how a forest area, destroyed by serious flooding by sea water, was colonized by *Euterpe chaunostachys*. We have only observed dense stands of *E. chaunostachys* in permanently waterlogged and periodically inundated areas which suggests that it can compete successfully only until land has built up by siltation and accumulation of organic material. Regeneration is found in the open (Lopez O. 1982) and in the shade of other trees but plants that grow in the open may flower earlier than plants in partial shade; in our study plot at Borbón we found one individual, fully exposed to the sun that had an inflorescence 1.8 meters above the ground but we have not observed flowering below five meters in dense stands.

Phenology
We frequently observed buds, inflorescences, and infructescences with green fruits of *Euterpe chaunostachys* during our fieldwork in northwestern Ecuador in February, 1988, but only once did we encounter mature fruits. According to local people the main fruiting season is March–April which coincides roughly with the time of maximum rainfall in the area (Climatic diagram 8 in Chapter 12) suggesting that the degree of flooding triggers the event. In Belém fruits of *E. oleracea*, collected in the Amazon estuary, are available all year round, but according to local merchants they come from different areas with distinct fruiting seasons (Strudwick and Sobel 1988). Distinct fruiting seasons in different areas of the Amazon was also noted by Wallace (1853).

Dispersal
Many seeds germinate below the mother tree, but since the palm is able to colonize it must also possess efficient means of dispersal. Water may act as dispersing agent, but this is probably the case only in areas with good current (as with *Astrocaryum jauari*), because fresh fruits and fresh and dry seeds will sink in water whereas dry fruits float for a while. According to local people in northwestern Ecuador, fruits of *E. chaunostachys* are consumed by a fish called "sabalo" which in Amazonian Ecuador is the spanish name for species of *Brycon* (Vickers 1976). Species of *Brycon* are known to consume fruits (Goulding 1983), and according to Ovchynnyk (1967) a number of *Brycon* species are present in the area (*Brycon acutus, B. atrocaudatus, B. dentex,* and *B. posadae*) and it is therefore likely that species of *Brycon* disperse seeds of *E. chaunostachys*. Goulding (1980) found that *Brycon* species act as predators as well as dispersers of fruits in the Amazon, and Patiño (1977) reported that fruits are consumed and dispersed by a species of large birds of the genus *Penelope,* and according to Snow (1981) fruits of *Euterpe* are consumed by specialist frugivorous birds.

Predation
Fish may be important predators on seeds, which suggests that *Euterpe*-seeds are important in the diet of these fish. Herbivorous crabs (*Gecarcinus lateralis,*

Cardisoma crassun) consume seedlings and newly germinated seeds of *Euterpe* (Lopez O. 1982).

Actual and potential uses

Inflorescences. — Pickles are made from inflorescences buds of various species of *Euterpe* (Uhl and Dransfield 1987). If *E. chaunostachys* buds can be used for this purpose they could easily be harvested along with the palm hearts; they could be canned, with the same equipment which is installed for canning palm hearts.

Fruits. — Fruits may be eaten raw (Barfod and Balslev 1988), but more commonly a beverage is made from them. Fruits as well as a beverage, *jugo de palmiche*, and ice cream made from the mesocarp of the fruits are sold in Borbón and San Lorenzo in coastal Ecuador. In Colombia the fruits are also used for beverages; the fruits are soaked and mashed in water to form a purple or, if the pericarp has been removed before mashing, white, fluid (Patiño 1977). The fresh pulp of *E. oleracea* contains 7–13% oil, 2.5–3.5% protein, 1–25% carbohydrates, and up to 18% fibers. As suggested for *Astrocaryum jauari*, fruits and seeds left over from beverage production could be used as fish fodder in ponds. In Brazil *E. oleracea* seeds are often fed to pigs (Strudwick and Sobel 1988).

Trunks. — Trunks are used for construction as house posts, roof poles, for fencing, and as pavements in swamps but only the older trunks are suitable for such purposes, the younger ones, though large enough to be cut for palm heart, being rather soft. The large number of trunks cut when harvesting palm hearts could possibly be used in the paper industry (CONIF 1980).

Leaves. — Leaves of *E. chaunostachys* are occasionally used for thatch, but other and more durable leaves are preferred when available. The palm heart is consumed locally and has become an export item. Palm heart is of little nutritional value (Quast and Bernhardt 1978), but because of its high protein content, on dry weight basis, it may play an underestimated role in the diet of indigenous people (Beckermann 1979, Table 11). The palm heart consists of young immature leaves in different development stages, hidden and protected by the leaf sheaths of the older leaves. Harvesting the palm heart kills the trunk (Fig. 20b).

Two canning factories operates in northwestern Ecuador. Both rely entirely on hearts from wild palms which are harvested by local people, who are paid according to the number of palm hearts they bring to the factories. The harvesters use canoes to get around in the swamps, often entering one of the many small tidal streams when the water level increases, leaving it again when the water level goes down. In order to harvest the heart, the trunk is cut and the outer leaf sheaths surrounding the heart are removed with a *machete,* but a few leaf sheaths are left to protect the heart during transport (Fig. 22). In this way it remains fresh for up to two days; removing all the sheaths would leave it

Figure 22. Harvest of palm hearts of *Euterpe chaunostachys* at Borbón on the Pacific coast of Ecuador. The crown-shaft is collected and the outer leaf sheaths are removed, leaving some of them to protect the soft palm heart while it is being transported to the canning factory; the canning process can wait for up until two days after harvest.

Figure 23. a. The canning of palm hearts of *Euterpe chaunostachys* at Borbón
involves the separation of the outer hard parts of the crownshaft from the inner soft parts.
b. The fresh palm hearts are placed in water with citric acid before they are canned.

useless in very short time because oxidation would make it brown and the delicate tissue would deteriorate.

Once at the factory the remaining leaf sheaths are removed (Fig. 23a, b). The canning process at Palmitto Borbón S. A. involves: 1. Cutting the heart into proper length, 2. Submerging the pieces into a mixture of water and citric acid, which helps to reduce browning, 3. Canning the pieces, together with water and citric acid which eliminates bacterial growth by reducing pH, 4. Following closure, the cans are placed in a bath with boiling water for about one hour, which sterilizes and softens the palm heart, 5. The cans are labeled and packed and are ready for sale. One palm heart, if not too young, is enough for one small can with a drained weight of 220 grams.

Waste products from the production, such as the leaf sheaths and the parts of the palm heart which are too hard and fibrous to be canned, are not used. Alcohol or animal fodder could be made from it (Lopez O. 1982), the latter possibility being investigated by Palmitto Borbón S. A. In Brazil waste products from canning of *E. oleracea* palm heart is used for mulch, producing a dark organic soil, and for animal ration, and leaves and infructescence stalks are used for mulch (Anderson 1988, Strudwick and Sobel 1988).

Management

Euterpe chaunostachys is occasionally grown near houses around Borbón in the Río Santiago area in Ecuador, but this is rare and the bulk of fruits and palm hearts are harvested from wild palms. The lack of adequate management practices for commercial harvest of palm hearts can have serious impacts on the stands; in Colombia *E. chaunostachys* was almost eliminated in some areas in the beginning of the 1970s, after which commercial harvest was prohibited for some years, but it has now been started again (CONIF 1980, Patiño 1977).

Two main types of management of natural stands may be employed: 1. Clear-cutting of all trunks with a palm heart large enough to be used, 2. Selective and continuous thinning of clusters, removing only a small number of trunks and leaving some to produce fruits. Clear-cutting may be sustainable if sufficient time for regeneration is allowed between harvests, but if large areas are clear cut, there will be a significant reduction of fruit production which may have a negative impact on fishery and local fruit consumption, and on long terms reduce or eliminate the regeneration of *E. chaunostachys*. Selective pruning of *E. oleracea* is common in the Amazon estuary where it usually is the young juvenile and old adult trunks that are cut, leaving the young adults and old juvenile. According to local people this practice increased fruit production on the remaining trunks. The management of *E. oleracea* stands in the Amazon estuary also involved thinning of other vegetation which was employed to increase fruit production and harvest of palm heart was of secondary importance (Anderson 1988). In Ecuador the situation is different; the fruits are not as much in demand as in Brazil and management is therefore directed towards securing a sustainable yield of palm hearts according to Palmitto Borbón S. A., and the harvesters are therefore told to leave at least one mature trunk in each cluster.

Table 11. Chemical composition of *Euterpe oleracea* palm heart in percent of dry weight except Vitamin C which is in Mg/100 g (From Quast and Bernhardt 1978).

	%
Protein	1.72
Ash	0.83
Crude fiber	0.27
Fat	0.08
Total sugars	0.70
Reducing sugars	0.30
Tannins	0.06
	mg/100 g
Vitamin C	1.4

To study the result of a harvest and to determine the density of *E. chaunostachys*, a 0.1 hectare study plot was made in an almost mono-specific natural stand near Río Santiago in Ecuador in a flat, swampy area which is almost completely inundated at high tide. Only trunks with a dbh over seven centimeters, corresponding to the minimum dbh of harvested trunks, were counted (Table 12). The plot included 37 clusters with at least one trunk with a dbh over seven centimeters; 65 trunks of a total of 125 had been cut, and 61 trunks were left, in 30% of the clusters no trunks with a dbh over seven centimeters were left, and in 43% more than one were left. Apart from the counted trunks each cluster had numerous shoots and seedlings and larger juvenile palms were numerous. Selection of trunks for cutting appeared to have been determined mainly by size, accessibility, and age; usually the oldest trunks were left because they are much harder to cut than the younger ones. We conclude that the stand will undoubtedly regenerate, given the large number of shoots, seedlings, and younger plants we observed, but the important question is when it will be harvested again. As more companies and many groups of harvesters operate in the area repeated harvest can happen any time and the selection method employed is likely to reduce fruit production significantly because mainly old trunks are left. Probably these are less productive than younger adults as found for *E. oleracea* (Anderson 1988).

Cultivation. — Cultivation of *E. chaunostachys* is hardly practiced in Ecuador although plantations have been planned but so far not established (García E. 1987). CONIF (1980) found good germination rates (77% in 90 days) and good survival rates (85%) of seedlings collected in the field and transplanted to nurseries in Colombia, and they established experimental plots in 1983, but so far growth is very poor. Absence of mycorrhiza may be the reason according to Enrique Vega (pers. comm.) which suggests that inoculation with mycorrhiza should be considered if cultivation is attempted outside the natural range of the palm. Planting densities can be high; the field data suggests that at least 370 clusters can be grown on one hectare, with a spacing of 5 x 5 meters, but no data are available on how early *E. chaunostachys* will start to flower, or when it will be large enough to be harvested for palm heart. The low-flowering individual that we found in the open (mentioned above) suggests that light conditions are important. In Brazil *Euterpe oleracea* reaches maturity in 3–4 years and farmers plant seedlings and seeds of *E. oleracea* in shifting cultivation plots in order to increase the usefulness of the areas during subsequent fallow periods (Anderson 1988).

Pests. — Seed predators may be important pests. All parts of *Euterpe oleracea* may be attacked by the aphid *Ceratephis lantania,* but it is not considered a serious pest. Larvae of the butterfly *Brassolis astyra* can consume large quantities of pinnae, and this may be a serious problem if not controlled (FAO 1987).

Production. — Trunks with a dbh over nine centimeters and trunks over four meters tall yield an average of 200 grams of palm heart (Lopez O. 1982). Little is known about fruit production. The only mature infructescence we found during our fieldwork had a weight of 1.6 kilograms, but, since it had already lost many fruits, this must be considered a minimum weight.

Harvest. — Occasionally flowering occurs as low as 1.8 meters above the ground and in such cases the fruits from young palms may be harvested from the ground with a *machete.* However this soon becomes impossible, and in Ecuador harvest of fruits and palm heart is normally done by felling the trunk. When *E. oleracea* fruits are harvested in the Amazon estuary it is a common practice to climb the trunk; a loop is made from a twisted palm leaf, the feet are placed in the loop and pressed tight to the trunk when climbing (Strudwick and Sobel 1988). This method is also employed to harvest palm heart, probably because it is easier to cut the crown-shaft than the trunk. Using a pole and a curved knife, fruits can be harvested from the ground because the infrafoliar infructescence is easy to reach and the peduncle is rather thin. Removal of fruits and palm hearts are not likely to have any significant influence on the nutrient balance in the areas. Usually *E. chaunostachys* grows in areas where regular flooding assures a continued input of nutrients.

Potential in different land-use systems

The areas where *E. chaunostachys* occurs have been relatively undisturbed until recent times but this is now changing. In Ecuador large scale commercial shrimp farming is spreading rapidly in coastal estuaries and mangroves (Jordan 1988) and in Colombia areas with *E. chaunostachys* are sometimes used for cultivation of rice (*Oryza sativa*) (Patiño 1977). These types of land-use can significantly reduce the possibilities of cultivation or management of natural stands of *Euterpe chaunostachys* within the natural range of the palm. So far nothing is known about the potential of the palm if grown on *terra firme.*

Table 12. Effect of palm heart harvest on a stand *Euterpe chaunostachys*. Data from a 0.1 hectare plot near Río Santiago in Ecuador. In the plot, 37 clusters had at least one trunk with a dbh over 7 cm. For each of these are given number of trunks with dbh over 7 centimeters, numbers of trunks cut and trunks left after harvest. Voucher specimen: *Pedersen no. 67301.*

| Cluster | Number of trunks with dbh > 7 cm | | |
	Total	Cut	Left
1	1	1	0
2	8	4	4
3	3	1	2
4	4	4	0
5	2	1	1
6	1	1	0
7	11	1	1
8	2	2	0
10	4	0	4
12	5	0	5
13	3	2	1
14	2	2	0
15	1	1	0
16	1	1	0
17	5	3	2
18	2	0	2
19	2	0	2
20	1	0	1
21	5	4	1
22	7	7	0
23	1	1	0
24	3	0	3
25	2	1	1
26	3	0	3
27	5	0	5
28	1	0	1
29	2	0	2
30	1	1	0
31	4	2	2
32	6	2	4
33	10	5	5
34	5	2	3
35	1	0	1
36	5	2	3
37	1	1	0
Total	126	65	61

Extractivism. — Euterpe chaunostachys is suitable for harvest from wild stands because its multi-stemmed habit acts as a buffer to over-exploitation and the continuous input of nutrients from inundations where it grows will replace nutrients lost with the crop. However, as mentioned above, there are examples that commercial extractivism can over-exploit the resource.

Agroforestry. — Euterpe chaunostachys is suitable for small scale mixed farming systems and it has several advantages in such systems. It yields cash crops such as fruits and palm heart, which can be consumed by the farmers themselves if market demands should fail and, if fish ponds are integrated in the farming system, the seeds may be used as fodder. The residues from fruit and palm heart production provide good mulch for other crop plants. Palm hearts can be harvested at any time of the year which means little interference with other farming activities, and the harvest can wait until enough palm hearts are mature to make transport to the canning factory worthwhile. The palm clusters can be shaped to fit the system; according to the light demands of other crops grown together with it, the palm clusters can be pruned more or less tightly. The superficial root system will help stabilize the swampy soils.

Plantations. — Euterpe chaunostachys is well adapted to grow in dense stands but the cost involved in establishing and maintaining a plantation is probably too high. Experience with palm heart plantations that use high yielding varieties of *Bactris gasipaes* has shown that competition from wild palms may be to high too favor plantations (Clement 1988), but more studies on yield, adaptability to different soils, effect of different management methods, fertilizer application, degree of pruning, and optimal harvest time, *etc.* are needed to give a meaningful evaluation of this possibility. Efficient utilization of fruits and "waste"-products would improve the economy of a plantation.

Conservation status

Given the taxonomic confusion regarding *Euterpe chaunostachys* it is difficult to determine its conservation status, but there is little doubt that its populations on the Pacific coast are threatened by over-exploitation, shrimp farming, and agriculture. We searched for *E. chaunostachys* on its type locality at El Recreo near Bahia de Caráquez in Ecuador, but did not find it; a large river valley nearby is completely occupied by shrimp farms ! In Colombia *E. cuatrecasana* was listed as endangered by the Conservation Monitoring Centre at Kew Gardens but transferring it to "vulnerable" has been suggested in Johnson (1986). In contrast to most other vulnerable species, which are threatened by destruction of their habitat, *E. cuatrecasana* is threatened by large scale destruction of its populations for the industrial production of palm heart (Bernal 1989). A previous report, however, estimated the area covered by this species to be 95,500 hectares which would indicate that it is not threatened (Tibaquira 1980).

   If *E. chaunostachys* turns out to be conspecific with the widespread *E. oleracea* the situation is different on the species level. However it is likely that the Pacific and

Amazonian populations, separated by the Andes, differ in a number of ways. Thus the threat to the Pacific *Euterpe*-populations is a threat to the genetic variation in the complex. The International Union for Conservation of Nature (IUCN) lists *Euterpe chaunostachys* as unknown, and *E. cuatrecasana* as vulnerable (Dransfield *et al.* 1988).

Research

Studies on regeneration of *E. chaunostachys* are carried out by Corporación Nacional de Investigación y Fomento Forestal (CONIF) in Colombia on the experimental station at San Isidro, near Buenaventura on the Pacific coast. Cultivation experiments based on seeds from Belém (*E. oleracea*) and from coastal Colombia was started in 1983 (Enrique Vega/ CONIF pers. comm.). The combination of *E. oleracea* and *E. chaunostachys* offers a possibility for taxonomic studies on the status of the two species. Leader of the cultivation experiments is Hoover Patiño.

Research on *E. oleracea* is done by INIPA in Iquitos, Peru, and by Centro de Pesquisa Agropecuaria dos Trópico Umido - CPATU (Coradin and Lleras 1988). Anthony B. Anderson from the Museo Paraense Emilio Goeldi has long been working on management of *E. oleracea* and is at present investigating the effect on different management methods (Anderson 1988). Andrew Henderson from the New York Botanical Garden and Gloria Galeano from Universidad Nacional in Bogotá have recently started a research project to resolve the taxonomic problems in *Euterpe*.

Listed in EMBRAPA / CENARGEN (1985) as researchers on *Euterpe* spp. are: Francis Kahn (Convenio Orstom/IIAP), working on systematic and ecology, E. Lleras Perez (IICA, S.A.I.N.), working on systematic, eco-physiology and domestication, and William L. Overall (Museo Paraense Emilio Goeldi) working on pollination.

Specimens studied

The following herbarium specimens were seen or collected for this study. For each specimen the following information is given: *Collector and number (in italics)*, locality, vegetation, elevation above sea level, date, and the acronyms of the herbaria where the specimens are deposited.

Bernal, R. and Galeano, G. no. 480. Colombia, Antioquia. Municipio de Mutatá. Carretera a Pavarandogrande. 100–150 m. 13 Dec 1982 (COL); *Bernal, R. and Galeano, G. no. 670.* Colombia, dep. de Antioquia. Municipio de Urrao. Corregimento de Vegáez. Bosque al Sur del Pueblo. 100–200 m. 16 Jul 1983 (COL, HUA); *Bernal, R. and Galeano, G. no. 891.* Colombia, dep. de Nariño. Municipio de Tumaco. Tangarial. 80 m. 4 Oct 1985 (AAU, COL); *Cuatrecasas no. 16901. Type.* Colombia, Intendencia (Departemento) del Chocó. Río San Juan, cercanias de Palestina. 12–14 Mar 1944 (Cited by Patiño 1977, not seen); *Fuchs, H. P., Zanella, L., and Torres R., J. H. no. 22053.* Colombia, dep. de Chocó. Region del Río Baudó, Lower Baudó area. 2 Feb–29 Mar 1967 (COL); *Eggers no. 15669. Type* Ecuador, El Recreo. Wald. 15 Feb 1897 (US); *Pedersen, H. B. no. 67301.* Ecuador, prov. Esmeraldas. Partly inundated swampforest near La Tola (79°05'W;1°10'N). The area dominated by *Euterpe* sp. 5 Feb 1988 (AAU, QCA).

12. CLIMATIC DIAGRAMS

Knowing the climatic conditions where palm species grow naturally, is important when attempting to bring them into cultivation or to exploit them in other ways. This section summarizes the most important climatic features in the general distribution area of the palm species treated in this publication with emphasis on the climatic conditions in Ecuador. The information is taken from Walter *et al.* (1975) and Cañadas Cruz (1983). The upper curve in the climatic diagrams gives the average monthly precipitation in millimeters according to the right hand scale; for precipitations over 100 millimeters per months the scale has been condensed to 10% of the scale below 100 millimeters. The lower curve is the average monthly temperature according to the left hand scale. The temperature scale and the precipitation scale are calibrated so that when the precipitation curve is above the temperature curve potential evapotranspiration is less than the precipitation which gives water surplus and wet climate. When the precipitation curve is below the temperature curve evapotranspiration is higher than precipitation and there is water deficit and a relatively dry climate. Water surplus is marked with vertical lines up to 100 millimeters precipitation per months and in full black over 100 millimeters. Water deficit, that is a dry season, is marked with a dotted signature. Over the diagram is given the name of the station, the elevation above sea level in round parenthesis ( ), the number of years or the time span for the data in square parenthesis [ ], the average annual temperature, and the average annual precipitation.

The climate map of South America shows a shift from high precipitation in the west to low precipitation in the east. It is also seen that the dry period north of equator is in the northern hemisphere winter, and vice versa south of the equator.

The climatic map of Ecuador shows general high precipitation in the areas where the palms included in this publication grow.

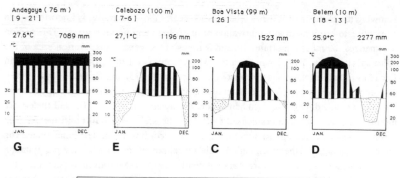

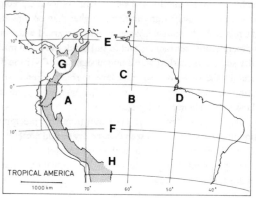

TROPICAL AMERICA

1000 km

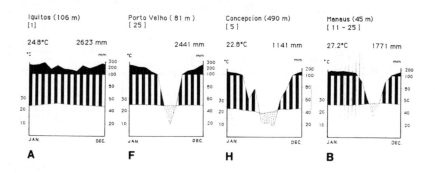

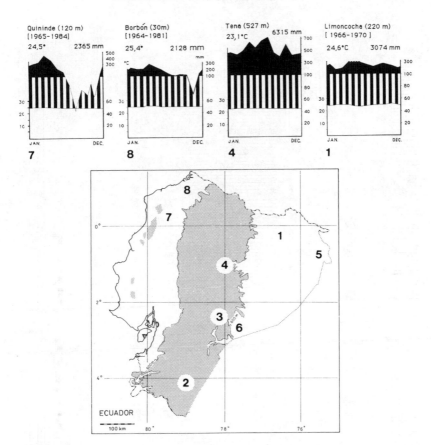

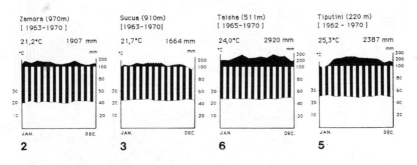

13. ADDRESSES

This section gives, in alphabetic order, the addresses of those institutions mentioned in the text where research is carried out on one or more of the species treated.

Aarhus University, Botanisk Institut — Herbariet, Bygn. 137, Universitetsparken, DK-8000 Aarhus C, Denmark.

CEPATU, Caixa Postal 48, Belém-PA 66 000, Brasil.

CONIF (Corporación Nacional de Investigacion y Fomento Forestal), Apartados Aéreos 095153 - 091676, Bogotá, D.E., Colombia.

Convenio Orstom/*II*AP, Apartado Postal 185, Iquitos, Peru.

CTAA / EMBRAPA (Centro de Tecnología de Alimentos) Av. das Américas, km. 295, Rio de Janeiro-RJ 23000, Brasil.

EMBRAPA / CENARGEN, S.A.I.N., Parque Rural, Caixa Postal 10-2372, Brasilia-DF 70770, Brasil.

Grupo Ultra, Av. Brigadeiro Luis Antonio 1343, São Paulo, SP 01317, Brasil.

Hoover Patiño, Apartado Aéreo 670, Buenaventura, Valle, Colombia.

ICRAF (International Council for Research in Agroforestry), P.O. Box 30677, Nairobi, Kenya.

IICA, S.A.I.N., Parque Rural, Caixa postal 10-2372, CEP 70.770, Brasilia-DF, Brasil

INIAP (Instituto Nacional de Investigaciones Agropecuarias), Administration central, Av. Eloy Alfaro y Amazonas. Edificio del MAG (4to. Piso), Telf. 567-645, Telex. 2532, Casilla 2600, Quito, Ecuador.

INIPA, Estación Experimental "Yurimaguas", Yurimaguas, Peru.

INPA, (Instituto Nacional de Pesquisas da Amazonia), 69000 Manaus, AM, Brasil.

International Technological Assistance B.V., P.O. Box 242, Winschoten, The Netherlands.

Museo Paraense Emilio Goeldi, Caixa Postal 399, Belém 66.000 Pará, Brasil.

New York Botanical Garden, Bronx, NY 10458, USA.

PIRB (Programa Interciencia de Recursos Biológicos) President: Dr. Alberto Ospina, Apartado Aéreo 51580, Bogotá, Colombia.

Universidad Nacional de Colombia: Facultad de Agronomia, Palmira, Valle, Colombia.

Universidad Nacional de Colombia, Instituto de Ciencias Naturales, Apartado Aéreo 7495, Bogotá, Colombia.

14. LITERATURE CITED

Acosta-Solís, M. 1944. La Tagua. — Editorial Ecuador, Quito.

Acosta-Solís, M. 1948. Tagua or vegetable ivory — A forest product of Ecuador. — Econ. Bot. 2: 46–57.

Acosta-Solís, M. 1952. Fibras y lanas vegetales en el Ecuador. — Contribución no. 21, Instituto Ecuatoriano de Ciencias Naturales. Quito.

Acosta-Solís, M. 1961. Los bosques del Ecuador y sus productos. — Editorial Ecuador, Quito.

Acosta-Solís, M. 1963. En Ecuador se estudian las palmas oleaginosas. — Reprint from "La Hacienda", October 1963.

Acosta-Solís, M. 1971. Palmas economicas del Noroccidente ecuatoriano. — Naturaleza Ecuatoriana 1(2): 80–163.

Alarcón G., R. 1988. Etnobotánica de los Quichuas de la Amazonía ecuatoriana. — Misc. Antrop. Ecuat. Ser. Monogr. 7: 1–183.

Anderson, A. B. 1978. The names and uses of palms among a tribe of Yanomama indians. — Principes 22: 30–41.

Anderson, A. B. 1988. Use and management of native forest dominated by Açaí palm (*Euterpe oleracea* Mart.) in the Amazon estuary. — Adv. Econ. Bot. 6: 144–154.

Anonymous 1986. Entrevistas. Debate sobre palma africana. — Colibrí 1: 47–64.

Arguëllo M., A. 1984. Estudio taxonómico preliminar de algunas especies representativas de las familias Arecaceae y Cyclanthaceae de la Reserva de la Corporación Forestal Juan Manuel Durini. — pp. 307–315 In: Memorias de la VIII Jornadas Nacionales de Ciencias Biológicas. — Universidad Tecnica de Ambato, Ecuador.

Arguëllo M., A. 1989. Arecaceae — pp. 53–64 In: Møller Jørgensen, P. and Ulloa U., C. (Eds.), Estudios botánicos en la "Reserva ENDESA" Pichincha, Ecuador. — AAU Reports 22.

Aristeguita, L. 1968. Consideraciones sobre la flora de los morichales llaneros al norte del Orinoco. — Acta Bot. Venz. 3: 19–38.

Balée, W. 1988. Indigenous adaptation to Amazonian palm forests. — Principes 32: 47–54.

Balick, M. J. 1976. The palm heart as a new commercial crop from tropical America. — Principes 20: 24–28.

Balick, M. J. 1979. Economic botany of the Guahibo. I. Palmae. — Econ. Bot. 33: 361–376.

Balick, M. J. 1981a. *Mauritiella* (Palmae) reconsidered. — Brittonia 33: 459–460.

Balick, M. J. 1981b. *Jessenia bataua* and *Oenocarpus* species; Native Amazonian palms as new sources of edible oil. — In: Pryde, E. H., Pincen, L. H. and Mukherjee, K. D. (Eds.), New sources of fats and oil. — American Oil Chemists Society, Champaign.

Balick, M. J. 1984. Ethnobotany of palms in the Neotropics. — Adv. Econ. Bot. 1: 9–23.

Balick, M. J. 1985. Current status of Amazonian oilpalms. — pp. 172–177 In: Pesce, C., Oilpalms and other oilseeds of the Amazon. Revised translation by Johnson, D. V. — Reference Publication Inc., Algonac, Michigan.

Balick, M. J. 1986. Systematics and economic botany of the *Oenocarpus-Jessenia* (Palmae) complex. — Adv. Econ. Bot. 3: 1–140.

Balick, M. J. 1988. *Jessenia* and *Oenocarpus*: Neotropical oil palms worthy of domestication. — FAO Plant Production and Protection Paper no. 88. FAO, Rome.

Balslev, H. 1987. Palmas nativas de la Amazonía ecuatoriana. — Colibrí 3: 64–73.

Balslev, H. and Barfod, A. 1987. Ecuadorean palms — an overview. — Opera Bot. 92: 17–35.

Balslev, H. and Henderson, A. 1986. *Elaeis oleifera* (Palmae) encontrada en el Ecuador. — Publ. Mus. Cienc. Nat. Ecuador 5: 45–49.

Balslev, H. and Henderson, A. 1987a. A new *Ammandra* (Palmae) from Ecuador. — Syst. Bot. 12: 501–504.

Balslev, H. and Henderson, A. 1987b. The identity of *Ynesa colenda* O. F. Cook (Palmae). — Brittonia 39: 1–6.

Balslev, H., Luteyn, J., Øllgaard, B., and Holm-Nielsen, L. B. 1987. Composition and structure of adjacent unflooded and floodplain forest in Amazonian Ecuador. — Opera Bot. 92: 37–57.

Balslev, H., Madsen, J., and Mix, R. 1988. Las plantas y el hombre en la Isla Puná, Ecuador. — Universidad Laica Vicente Rocafuerte de Guayaquil y Museo Antropológico, Banco Central del Ecuador, Guayaquil.

Barbosa Rodrigues, J. 1903a. Sertum palmarum brasiliensium. — Veuve Monnom, Brussels.

Barbosa Rodrigues, J. 1903b. Les noches des palmiers. — Imprimerie de Mertens, Brussells.

Barfod, A. 1988. Natural history of the subfamily *Phytelephantoideae* (Arecaceae). — PhD thesis, Botanical Institute, Aarhus University, Denmark.

Barfod, A. and Balslev, H. 1988. The use of palms by the Cayapas and Coaiqueres on the coastal plain of Ecuador. — Principes 32: 29–42.

Barfod, A., Henderson, A., and Balslev, H. 1987. A note on the pollination of *Phytelephas microcarpa* (Palmae). — Biotropica 19: 191–192.

Barfod, A., Bergmann, B., and Pedersen, H. B. (in press) The vegetable ivory has survived and is doing fine in Ecuador. — Econ. Bot.

Barrett, S. A. 1925. The Cayapa indians of Ecuador. — Indians Notes and Monog., Museum of the American Indians, Heye Foundation, New York 40: 1–476.

Bavappa, K. V. A., Nair, M. K., and Prem Kumar, T. 1982. The arecanut palm. — Mathrubhumi Press, Calicut, India.

Beccari, O. 1916. Il Genera *Cocos* Linn. e le palme affine. — Agric. Colon 10(2): 435–471.

Beckerman, S. 1979. The abundance of protein in Amazonia: A reply to Gross. — Amer. Anthropol. 81: 533–560.

Bene, J. G., Beall, H. W., and Cote, A. 1977. Food and people: Land management in the tropics. — IDRC, Ottawa.

Bernal, R. G. 1989. Endagerment of Colombian palms. — Principes 33: 113–128.

Berry, P. 1976. Estudio bibliografico y taxonomico preliminar sobre palma "Seje". — Report, CODESUR, Caracas.

Bianchi, C., Rovere, F., Clemente, T., Brosegehini, S., Palacios-Galo Espinosa, A., Fruci, S., and Bottasso, J. 1982. Artesanias y tecnicas Shuar. — Mundo Shuar, Quito.

Bishop, J. B. 1980. Agroforestry systems for the humid tropics east of the Andes. — Report, INIAP, Quito.

Blaak, G. 1988. Mechanical extraction and prospects for development of a rural industry. — In: Balick, M. J. (Ed.), *Jessenia* and *Oenocarpus*: Neotropical oil palms worthy of domestication. — FAO Plant Production and Protection Paper no. 88. FAO, Rome.

Blicher-Mathiesen, U. and Balslev, H. (in press). *Attalea colenda* (Palmae), a potential lauric oil resource. — Econ. Bot.

Blydenstein, J. 1967. Tropical savanna vegetation of the llanos of Colombia. — Ecology 48: 1–15.

Bodley, J. H. and Benson, C. F. 1979. Cultural ecology of Amazonian palms. — Reports of Investigations 56, Laboratory of Anthropology, Washington State University, Pullman.

Bohórquez R., J. A. 1976. Monografia sobre *Mauritia flexuosa* L.f — pp. 233–244 In: IICA (Ed.), Simposio Internacional sobre Plantas de Interes Economico de la

Flora Amazonica, Belém, Brasil,1972. — Turrialba.
Boldt, K. 1982. Kokos, et naturprodukt. — Skarv, Holte.
Boom, B. M. 1986. The Chacobo indians and their palms. — Principes 30: 63–70.
Borchsenius, F. and Balslev, H. 1990. Three new species of *Aiphanes* and notes on the genus in Ecuador. — Nord. J. Bot. 9: 383–393.
Braun, A. 1968. Cultivated palms of Venezuela. — Reprint of Principes 12 (2,3,4).
Braun, A. and Delascio C., F. 1987. palmas autóctonas de Venezuela y de los países adyacentes. — Caracas.
Bullock, S. H. 1981. Notes on the phenology of inflorescences and pollination of some rainforest palms in Costa Rica. — Principes 25: 101–105.
Burley, J. and Carlowitch, P., von 1984. Definition of multipurpose trees. — pp. 1–4 In: Burley, J. and Carlowitch, P., von (Eds.), Multipurpose tree germplasm. — ICRAF, Nairobi.
Burret, M. 1928. Die Palmengattung *Chelyocarpus* Dammer und *Tessmanniophoenix* Burret nov. gen. — Notizbl. Bot. Gart. Berlin-Dahlem 10: 394–401.
Burret, M. 1929a. Die Gattung *Ceroxylon* Humb. et. Bonpl. — Notizbl. Bot. Gart. Berlin-Dahlem 10: 841–854.
Burret, M. 1929b. Die Gattung *Hyospathe* Mart. — Notizbl. Bot. Gart. Berlin-Dahlem 10: 854–859.
Burret, M. 1929c. Die Gattung *Euterpe* Gaertn. — Bot. Jahrb. Syst. 63: 49–76.
Burret, M. 1930. Eine neue Palmengattung aus Südamerika. — Notizbl. Bot. Gart. Berlin-Dahlem 11: 48–51.
Burret, M. 1931. Palmae Hoppianae novae vel criticae. — Notizbl. Bot. Gart. Berlin-Dahlem 11: 231–236.
Burret, M. 1932. Die Palmengattung *Martinezia* und *Aiphanes*. — Notizbl. Bot. Gart. Berlin-Dahlem 11: 557–577.
Burret, M. 1933. *Chamaedorea* Willd. und verwandte Palmengattungen. — Notizbl. Bot. Gart. Berlin-Dahlem 11: 724–768.
Burret, M. 1934. Die Palmengattung *Astrocaryum* G. F. W. Meyer. — Repert. Spec. Nov. Regni. Veg. 35: 114–158.
Burret, M. 1936a. Die Palmengattung *Morenia* R. and P. — Notizbl. Bot. Gart. Berlin-Dahlem 13: 332–339.
Burret, M. 1936b. Palmae Neogeae X. — Notizbl. Bot. Gart. Berlin-Dahlem 13: 339–347.
Burret, M. 1937. Die Palmengattung *Syagrus* Mart. — Notizbl. Bot. Gart. Berlin-Dahlem 13: 677–696.
Burret, M. 1939. Palmae. — In: Diels, L. Neue Arten aus Ecuador II. — Notizbl. Bot. Gart. Berlin-Dahlem 14: 324–329.
Burret, M. 1940. Palmae. — In: Diels, L. Neue Arten aus Ecuador III. — Notizbl. Bot. Gart. Berlin-Dahlem 15: 23–38.
Calzada, B. J. 1980. 143 Frutales nativos. — Publicación de la Universidad Nacional Agraria La Molina, La Molina, Peru.
Cañadas Cruz L. 1983. El mapa bioclimatico y ecologico del Ecuador. — MAG-PRONAREG, Quito.
Carrión, J. 1970. Extración de aceite de Chapil (*Jessenia polycarpa*). — Thesis, Ingenería Quimica, Escuela Politecnica Nacional, Quito.
Carrión, L. and Cuvi, M. 1985. La palma africana en el Ecuador: Tecnologia y expanción empresarial. — Facultad Latinoamericano de Ciencias Sociales (FLACSO), Quito.
Cavalcante, P. B. 1974. Frutas comestiveis da Amazônia (2). — Museo Paraense Emílio Goeldi, Belém.
Civrieux, J. M. S. De 1957. Nombres folklóricos y indigenas de algunas palmeras Amazonico-Guayanesas con apuntos etnobotánicos. — Bol. Soc. Venz. Cien. Nat.

18(89): 195–233.

Clement, C. R. 1986. The Pejibaye palm (*Bactris gasipaes* H.B.K.) as an agroforestry component. — Agroforestry Systems 4: 205–219.

Clement, C. R. 1988. Domestication of the Pejibaye palm (*Bactris gasipaes*): Past and present. — Adv. Econ. Bot. 6: 155–174.

Clement, C. R. and Mora Urpi, J. 1987. Pejibaye palm (*Bactris gasipaes*, Arecaceae): Multi-use potential for the lowland humid tropics. — Econ. Bot. 41: 302–311.

CONFENIAE 1985. palma africana y etnocidio. — CEDIS, Quito.

Collazos T., M. E. 1987. Fenologia y postcosecha de Mil Pesos, *Jessenia bataua* (Mart.) Burret. — Thesis, Universidad Nacional de Colombia, Facultad de Ciencias Agropecuarias, palmira.

CONIF 1980. Informe de las actividades agroforestales de CONIF y el projecto INDERENA-FAO en San Isidro (Municip. de Buenaventura). — Report.

Cook, O. F. 1942. A new commercial oil palm in Ecuador. — Natl. Hort. Mag. april 1942: 70–85.

Coradin, L. and Lleras, E. 1988. Overview of palm domestication in Latin America. — Adv. Econ. Bot. 6: 175–189.

Corner, E. J. H. 1966. The natural history of palms. — Weidenfeld and Nicolson, London.

Dahlgren, B. E. 1936. Index of American palms. — Field Mus. Nat. Hist. Bot. Ser. 14: 1–456.

Dahlgren, B. E. 1944. Economic products of palms. — Trop. Woods 78: 10–35.

Danida 1988. Handlingsplan: Miljø og udvikling. — Danida, København.

Davis, T. A. 1984. A climbing devise to reach the palmyra's valuable "toddy". — Spirit of Enterprise, The 1984 Rolex Awards. — Arum Press, London.

De los Heros G., M. J. and Zárate, J. B. 1980–1981. Possibilidades papeleras de pulpa al sulfato de peciolos de aguaje. — Rev. Forestal del Peru 10: 83–90.

Dodson, G. H. and Gentry, A. H. 1978. Flora of the Río Palenque Science Center. — Selbyana 4: 1–628.

Dransfield, J., Johnson, D. V., and Synge, H. 1988. The palms of the New World: A conservation census. — IUCN, Gland, Switzerland and Cambridge, UK.

Dugand, A. 1940. Palmas de Colombia: Clave diagnóstica de los generos y nómina de las especies conocidas. — Caldasia 1: 20–84.

Dugand, A. 1972. Las palmeras y el hombre. — Cespedesia 1(1–2): 31–103.

Dugand, A. 1976. Palmarum colombiensum elenchus. — Cespedesia 5(19–20): 257–326.

Eckey, E. W. 1954. Vegetable fats and oils. — Reinhold Publishing Corporation, New York.

EMBRAPA/CENARGEN 1985. Directory of researchers in Neotropical palms. — Newsletter, Useful palms of Tropical America 1: 12–13.

EMBRAPA/CENARGEN 1986. Directory of researchers on Neotropical palms. — Newsletter, Useful palms of Tropical America 2: 13.

Estrada, R. O., Seré, C., and Luzuriaga, H. 1988. Sistemas de producción agrosilvopastorales en la selva baja de la provincia del Napo, Ecuador. — CIAT, Cali.

Estrella, E. 1988. El pan de America. — Ediciones Abya-Yala, Quito.

FAO 1986. La madera de Coco. Elaboración y aprovechamiento. — Estudio FAO Montes 57. FAO, Rome.

FAO 1987. Especies forestales productoras de frutas y otros alimentos. — Estudio FAO Montes 44/3. FAO, Rome.

FAO/CATIE 1984. Palmeras poco utilizadas de America Tropical. — LIL, S.A., Turrialba.

Ferwerda, J. D. 1984. Oil crops. — pp. 59–108 In: Martin, F. W. (Ed.), Handbook

of Tropical Foodcrops. — CRC Press Inc., Boca Raton.
Forero, L. E. 1983. Anotaciones sobre bibliografia seleccionada de complejo *Jessenia-Oenocarpus* (Palmae). — Cespedesia 45–46: 21–43.
Franke, G. 1982. Nutzpflanzen der Tropen und Subtropen, ed. 4. — S. Hirzel, Leipzig.
Galeano, G. and Bernal, R. 1987. Palmas del Departamento de Antioquia. — Universidad Nacional de Colombia, Bogotá.
García B., F. 1974. Plantas medicinales de Colombia. — Museo de Ciencias Naturales, Univ. Nacional, Bogotá.
García E., J. 1987. Estudio de forestación de palmitos en la Republica del Ecuador. — Report, HVA-International B. V., Amsterdam.
García S., M. 1986. Palmas nativas nueva fuenta de alimento. — Colibrí 1: 65–68.
García S., M. 1988. Observaciones de polinización en *Jessenia bataua* (Arecaceae). — Thesis de licenciatura, P. Univ. Católica del Ecuador, Dep. de Ciencias Biológicas, Quito.
Glassman, S. F. 1965. Geographic distribution of new world palms. — Principes 9: 132–134.
Glassman, S. F. 1970. A synopsis of the palm genus *Syagrus* Mart. — Fieldiana, Bot. 32: 215–241.
Glassman, S. F. 1972. A revision of B. E. Dahlgren's index of American palms. — Phanerogamarum Monographiae, Tomus VI. — Cramer, Germany.
Gonzáles, R.M. 1971–1974. Estudio sobre la densidad de poblaciones de aguaje (*Mauritia* sp.) en Tingo Maria, Peru. — Report.
Goulding, M. 1980. The fishes and the forest. — University of California Press, Berkely.
Goulding, M. 1981. Man and fisheries on an Amazon Frontier. — Dr. W. Junk Publishers, The Hague.
Goulding, M. 1983. The role of fishes in seed dispersal and plant distribution in Amazonian floodplain ecosystems. — pp. 271–283 In: K. Kubitzki (Ed.), Dispersal and distribution, an international symposium. — Sonderbd. Naturwiss. Ver. Hamburg 7.
Granville, J. J. De. 1974. Aperçu sur la structure des pneumatophores de deux espèces des sols hydromorphes en Guyane: *Mauritia flexuosa* L. et *Euterpe oleracea* Mart. (Palmae). Generalisation au système respiratoire racinaire d´autres palmiers. — Cahiers ORSTOM. Series Biologie. 23: 3–22.
Grimwood, B. E. 1975. Coconut palm products. — FAO, Rome.
Hartley, C. W. S. 1977. The oil palm, ed. 2. — Longman, London.
Hecht, S. 1983. Cattle ranching in the eastern Amazon; Environmental and social implications. — pp. 155–188 In: Moran, E. F. (Ed.), The dilemma of Amazonian development. — Westview Press, Boulder.
Heinen, H. D. and Ruddle, K. 1974. Ecology, ritual, and economic organization in the distribution of palm starch among the Warao of the Orinoco delta. — J. Anthrop. Research 30: 116–138.
Henderson, A. 1986. A review of pollination studies in the Palmae. — Bot. Rev. 52: 221–259.
Henderson, A. and Balick, M. 1987. Notes on the palms of Amazônia Legal. — Principes 31: 116–122.
Huber, J. 1906. La Végétation de la vallé du Rio Purus (Amazone). — Bull. L´Herbier Boissier 5: 249–275.
Iglesias, G. 1985. Hierbas medicinales de los Quichuas del Napo. — Abya-Yala, Quito.
Jamieson, G. S. and McKinney, R. S. 1934. Patua palm oil. — Oil and Soap 11: 207, 217–218.
Jijón, C. 1986. Palma africana ¿Deterio ecológico o social?. — Colibrí 1: 37–43.
Johnson, D. V. 1983. Multi-purpose palms in agroforestry: A classification and

assessment. — Int. Tree Crops J. 2: 217–244.
Johnson, D. V. 1985. Brasilian oilseed production by state, 1980. — pp. 185–187 In: Pesce, C. 1985. Oil palms and other oilseeds of the Amazon. Revised translation by Johnson, D. V. from Oleaginosas da Amazonia. — Reference Publications, Inc., Algonac, Michigan.
Johnson, D. V. 1986. Economic botany and threatened species of the palm family in Latin America and Carribbean. — Final Report WWF 3322 (Palms).
Johnson, D. V. 1988. Worldwide endangerment of useful palms. — Adv. Econ. Bot. 6: 268–273.
Jordan, C. B. 1970. A Study of germination and use in twelve palms of northeastern Peru. — Principes 14: 26–32.
Jordan, E. 1988. Die Mangrovenvälder Ecuadors im Spannungsfeld zwischen Ökologie and Ökonomie. — Jahrbuch der Geographischen Gesselschaft zu Hannover. — Selbstverlag der Geographischen Gesselschaft, Hannover.
Junk, W. J. 1989. Flood tolerance and tree distribution in central Amazonian floodplains. — pp. 47–64 In: Holm-Nielsen, L. B., Nielsen, I. C., and Balslev, H. (Eds.), Tropical forests: Botanical dynamics, speciation, and diversity. — Academic Press, London.
Junk, W. J. and Furch, K. 1985. The physical and chemical properties of Amazonian waters and their relationship with the biota. — pp. 3–17 In: Prance, G. T. and Lovejoy, T. E. (Eds.), Amazonia. — Pergamon Press, Oxford.
Kahn, F. 1988. Ecology of economically important palms in Peruvian Amazon. — Adv. Econ. Bot. 6: 42–49.
Kahn, F. 1990. Las palmeras del Arboretum Jenaro Herrera (Provincia de Requena, Departamento de Loreto, Peru). — Candollea 45: 341–362.
Kiltie, R. A. 1981. Stomach contents of rainforest peccaries (*Tayassu tajacu* and *T. pecari*). — Biotropica 13: 234–236.
Korning, J. and Thomsen, K. 1988. Studies of Amazonian tree and understory vegetation and associated soils in Añango, east Ecuador. — C.Sc. Thesis, Botanical Institute, Aarhus University, Denmark..
Kraus, G. 1896. Physiologisches aus den Tropen. — Ann. Jardin Bot. Biutenzorg 13: 217–275.
Kvist, L. P. and Holm-Nielsen, L. B. 1987. Ethnobotanical aspects of lowland Ecuador. — Opera Bot. 92: 83–107.
Lamb, B. 1959. The coastal swamp forest of Nariño Colombia. — Carib. For. 20: 79–98.
Lescure, J. P., Balslev, H., and Alarcón, R. 1987. Plantas útiles de la Amazonia Ecuatoriana. — ORSTOM, PUCE, INCRAE and PRONAREG, Quito.
Leslie, A. J. 1987. A second look at the economics of natural management systems in tropical mixed forests. — Unasylva 39: 46–58.
Lévi-Strauss, C. 1952. The use of wild plants in tropical South America. — Econ. Bot. 6: 252–270.
Little, E. L., Jr. 1970. New tree species from Esmeraldas, Ecuador. — Phytologia 19: 251–259.
Little, E. L., Jr. and Dixon, R. G. 1969. Arboles comunes de la Provincia de Esmeraldas. — FAO (FAO/SF: 76/ECU 13), Rome.
Lleras, E. and Coradin, L. 1988. Native Neotropical oil palms: State of the art and perspectives for Latin America. — Adv. Econ. Bot. 6: 201–213.
Lopez C., R. 1968. Ensayos de germinación de *Mauritia flexuosa* L.f. — Univ. Nacional de la Amazonía, Iquitos.
Lopez O., G. E. 1982. Plan de ordenacion forestal (*Euterpe* spp.). — Latinoamericano de Industria y Comercio S. A. (LICSA), Bogotá.
MacBride, J. F. 1960. Palmae in Flora of Peru. — Field Mus. Nat. Hist., Bot. Ser. 13: 321–418.

MAG 1975. El Cocotero y su cultivo. Manual Tecnico no. 1-DA-5-75. — Ministerio de Agricultura y Ganaderia, Quito.

Markley, K. S. 1949. FAO oilseed mission for Venezuela. — FAO, Washington.

Marles, R. J., Neill, D. A., and Farnsworth, N. R. 1988. A contribution to the ethnopharmacology of the lowland Quichua people of Amazonian Ecuador. — Revista de la Academia Colombiana de Ciencias Exactas, Fisicas y Naturales 16 (63): 111–120.

Martius, C. F. P. von 1823. Historia naturalis palmarum 2: 23. Munich.

May, P. H., Anderson, A. B., Balick, M. J., and Frazão, J. M. F. 1985. Subsistence benefits from the Babassu palm (*Orbignya martiana*). — Econ. Bot. 39: 113–129.

Mazzani, B., Oropeza, H., and Malaguti, G. 1975. El Seje. — Coco y palma 10.

Mejia C., K. 1988. Utilization of palms in eleven mestizo villages of the Peruvian Amazon (Ucayali River, Department of Loreto). — Adv. Econ. Bot. 6: 130–136.

Meunier, J. and Hardon, J. J. 1976. Interspecific hybrids between *Elaeis guineensis* and *Elaeis oleifera*. — pp. 127–138 In: Corley, R. H. V., Hardon, J. J., and Wood, B. J. (Eds.), Oil palm research. — Elsevier, Amsterdam.

Moldenke, H. N. 1949. The botanical source of patava oil. — Phytologia 3: 122–129.

Moraes R., M. and Henderson, A. 1990. The genus *Parajubaea* (Palmae). — Brittonia (in press).

Moraes R., M. 1989. Ecología y formas de vida de las palmas Bolivianas. — Ecología en Bolivia 13: 33–45.

Moore, H. E., Jr. 1973a. The major groups of palms and their distribution. — Gentes Herb. 11: 27–141.

Moore, H. E., Jr. 1973b. Palms in the tropical forest ecosystems of Africa and South America. — pp. 63–88 In: Meggers, B. J., Ayensu, E. S., and Duckworth, W. D. (Eds.), Tropical forest ecosystems in Africa and South America: A comparative review. — Smithsonian Institution Press, Washington, D. C.

Moore, H. E., Jr. 1979. Endangerment at the specific and generic levels in palms. — Principes 23: 47–64.

Moore, H. E., Jr. 1980. Four new species of Palmae from South America. — Gentes Herb. 12: 30–38.

Moore, H. E., Jr. 1982. A new species of *Wettinia* (Palmae) from Ecuador. — Principes 26: 42–43.

Mora Urpi, J., Vargas, E., López, C. A., Villaplana, M., Allón, G., and Blasco, C. 1982. El Pejibaye. — Publicación del Banco Nacional de Costa Rica, San José.

Nair, P. K. R. 1979. Intensive multiple cropping with coconuts in India: Principles, programmes and prospects. — Parey, Berlin, Hamburg.

Nair, P. K. R. 1980. Agroforestry species — A crop sheets manual. — ICRAF, Nairobi, pp. 3–47.

Nodari, R. O. and Guerra, M. P. 1986. O palmiteiro no sul do Brazil: Situação e perspectivas. — Useful palms of Tropical America 2: 9–10.

Nogueira, J. B. and Machado, D. R. 1950. Glossário de palmeiras. Oleaginosas e Ceríferas. — Instituto de ´Oleos, Ministerio da Agricultura, Rio de Janeiro.

ONERN 1977. Use of remote sensing systems. Evaluating the potential of the Aguaje palm tree in the Peruvian jungle. — Report, ONERN, Lima.

Orellana M., F. 1986. Control biológico del insecto defoliador de la palma africana *Sibine fusca* Stoll (Lepidoptera-Limacodidae). — Bol. Divulgativo (INIAP-Ecuador) 170.

Orr, C. and Wrisley, B. 1981. Vocabulario de Quichua del Oriente. ed. 2. — Inst. Ling. de Verano, Quito.

Ovchynnyk, M. M. 1967. Freshwater fishes of Ecuador. — Michigan State University, East Lansing.

Pacheco, F. P. 1989. Caracterización fisico-quimico del morete (*Mauritia flexuosa*).

— Thesis, Ingeneria Quimico, Escuela Politecnica Nacional, Quito.

Padoch, C. 1988. Aguaje (*Mauritia flexuosa* L. f.) in the economy of Iquitos, Peru. — Adv. Econ. Bot. 6: 214–224.

Patiño, V. M. 1977. Palmas oleaginosas de la costa colombiana del Pacifico. — Cespedesia 6(23–24): 131–260.

Pellizaro, S. M. 1978. La celebración de Uwi. — Publicación de los Museos del Banco Central del Ecuador, Quito-Guayaquil.

Perez-Arbelaez, E. 1978. Plantas útiles de Colombia. — Litografia Arco, Bogotá.

Pesce, C. 1985. Oil palms and other oilseeds of the Amazon. Revised translation by Johnson, D. V. of "Oleaginosas da Amazonia." — Reference Publications, Inc., Algonac, Michigan.

Piedade, M. T. F. 1985. Ecologia e biologia reprodutiva de *Astrocaryum jauari* Mart. (Palmae) como exemplo de população adaptada a áreas inundaveis do Rio Negro (Igapós). — M. Sc. Thesis. INPA/ Fundação Universidad do Amazonas, Manaus .

Pinheiro, C. U. B. and Balick, M. J. 1987. Brazilian palms. Notes on their uses and vernacular names, compiled and translated from Pio Corrêa's "Dictionário das Plantas Uteis do Brazil e das Exóticas Cultivadas," with updated nomenclature and added illustrations. — Contr. New York Bot. Gard. 17: 1–63.

Plotkin, M. J. and Balick, M. J. 1984. Medicinal uses of South American palms. — J. Ethnopharmacology 10: 157–179.

Posey, D. A. 1984. A preliminary report on diversified management of tropical forest by the Kayapó indians of the Brazilian Amazon. — Adv. Econ. Bot. 1: 112–126.

Poulsen, G. 1978. Man and tree in tropical Africa. — IDRC Report 101e. IDRC, Ottawa.

Prance, G. T. 1979. Notes on the vegetation of Amazonia III. The terminology of Amazonian forest types subject to inundations. — Brittonia 31: 26–28.

Quast, D. G. and Bernhardt, L. W. 1978. Progress in palmitto (heart of palm) processing research. — J. Food Protection 41(8): 667–674.

Raintree, J. B. 1987. The state of the art of agroforestry diagnosis and design. — Agroforestry Systems 5: 219–250.

Ranghel G., A. 1945. La palmera milpés o seje de la Amazonia Colombiana. — Agricultura Tropical 6: 40–43.

Raulino R., P. 1974. Palmeiras, Flora Ilustrada Catarinense. — Itajaí, Santa Catarina.

Richards, P. W. 1979. The tropical rainforest, ed. 7. — University Press, Cambridge.

Ridley, H. N. 1930. The dispersal of plants throughout the World. — L. Reeve and Co., Ltd., Ashford.

Rivadeneira Z., J. 1985. Fertilización mineral de palma africana en el sitio definitivo. — Bol. Divulgativo (INIAP-Ecuador) 164.

Ruddle, K., Johnson, D., Townsend, K. P., and Rees, J. D. 1978. Palm sago, a tropical starch from marginal lands, 1. ed. — The University Press of Hawaii, Honolulu.

Salazar C., A. 1967. El aguaje (*Mauritia vinifera*) recurso forestal potencial. — Rev. Forestal del Peru 1: 65–68.

Salazar C., A. and Roessl, J. 1977. Estudio de la potencialidad industrial del Aguaje. — Proyecto ITINTEC 3102 UNA-IIA, Lima.

Sánchez-Monge y Parellada, E. 1980. Diccionario de plantas agrícolas. — Ministerio de Agricultura, Servicio de Publicaciones Agrarias, Madrid.

Sarmiento, G. 1983. The savannas of tropical Americas. — pp. 245–288 In: Bourlière, F. (Ed.), Tropical Savannas, Ecosystem of the world 13. — Elsevier Scientific Publishing Company, Amsterdam.

Schlüter, U. 1989. Morphologische, anatomische und physiologische Untersuchungen zur überflutungstoleranz zweier charakteristischer Baumarten (*Astrocaryum jauari* und *Macrolobium acaciaefolium*) des Weiss- und Schwarzwasser-überschwemmungswaldes bei Manaus. — Ein beitrag zur Ökosystemanalyse von

Várzea und Igapó Zentralamazoniens.— Ph. D. Thesis, Kiel University, FRG.

Schultes, R. E. 1974. Palms and religion in the northwest Amazon. — Principes 18: 3–21.

Schultes, R. E. 1977. Promising structural fiber palms of the Colombian Amazon. — Principes 21: 72–82.

Schønheyder, F. and Nørby, J. 1965. Biokemi, ed. 4. — Universitetsforlaget, Aarhus.

Sirotty, L. and Malagotty, G. 1950. La agricultura en el Territorio Amazonas: Explotación del seje (*Jessenia bataua*) palma oleaginosa. — Caracas.

Sist, P. and Puig, H. 1987. Régéneration, dynamique des populations et dissémination d´un palmier de Guyane Française: *Jessenia bataua* (Mart.) Burret subsp. *oligocarpa* (Griseb, and H. Wendl) Balick. — Bull. Mus. natn. Hist. nat., Paris, 4é sér., 9, section B, Adansonia 3: 317–336.

Skov, F. and Balslev, H. 1989. A revision of *Hyospathe* (Palmae). — Nord. J. Bot. 9: 189–202.

Snow, D. W. 1981. Tropical frugivorous birds and their food plants: A world survey. — Biotropica 13: 1–14.

Spruce, R. 1908 [1970]. Notes of a botanist on the Amazon and Andes (1st. ed. 1908). — Johnson Reprint Corporation, New York.

St. John, T. V. 1988. Prospects for application of vesicular-arbuscular mycorrhizae in the culture of tropical palms. — Adv. Econ. Bot. 6: 50–55.

Strudwick, J. and Sobel, G.L. 1988. Uses of *Euterpe oleracea* Mart. in the Amazon estuary, Brazil. — Adv. Econ. Bot. 6: 225–253.

Tibaquira C., L. 1980. Potencial de los bosques de palma Naidi en la costa sur del Pacifico Colombiana, (Cauca, Nariño). — Mimeograph, INDERENA, Bogotá.

Timell, T. E. 1957. Vegetable ivory as a source of a mannan polysaccharide. — Canad. J. Chem, 35: 333–338.

Uhl, N. W. and Dransfield, J. 1987. Genera palmarum. — Allan Press, Lawrence.

Urrego G., L. E. 1987. Estudio preliminar de la fenologia de la Cananguchi (*Mauritia flexuosa* L. f.). — Tesis, Universidad Nacional del Colombia, Facultad de Agronomia. Medellin.

Vera D., H. and Orellana M., F. 1986. Sagalassa valida, el "Gusano Barrenador" de las raices de la palma africana y su combata. — Bol. Divulgativo (INIAP-Ecuador) 190.

Vickers, W. T. 1976. Cultural adaption to Amazonian habitats: The Siona-Secoya of Eastern Ecuador. — PhD thesis, University of Florida, Gainesville.

Vickers, W. T. 1978. Native Amazonian subsistence in diverse habitats: The Siona-Secoya indians of Ecuador. — pp. 6–36 In: Zamorra, M. D., Sutlive, V. H., and Altshuler, N. (Eds.), Changing agricultural systems in Latinamerica. — Studies in Third World Communities. Publ. no. 7. Dep. Anthrop., Coll. of William and Mary, Williamsburg.

Vickers, W. T. and Plowman, T. 1984. Useful plants of the Siona and Secoya indians of eastern Ecuador. — Fieldiana, Bot. 15: 1–63.

Wallace, A. R. 1853. Palm trees of the Amazon and their uses. — John van Vooerst, London.

Walter, H., Harnickell E., and Mueller-Dombois, D. 1975. Klimadiagramm-Karten. — Gustav Ficher Verlag, Stuttgart.

Wessels Boer, J. G. 1965. The indigenous palms of Suriname.— E. J. Brill, Leiden.

Wessels Boer, J. G. 1968. The geonomoid palms. — Verh. Kon Ned. Akad. Wetensch., Afd. Natuurk., Tweede Sect. 58(1): 1–202.

Wheeler, M. A. 1970. Siona use of Chambira palm fiber. — Econ. Bot. 24: 180–181.

15. INDEX TO SCIENTIFIC NAMES